REGINALD FARRER

REGINALD FARRER

Dalesman, planthunter, gardener

edited by

John Illingworth and **Jane Routh**

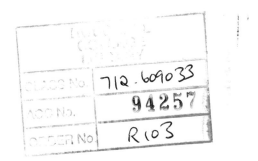
Centre for North-West Regional Studies
University of Lancaster

Occasional Paper no. 19

General editor: Oliver M. Westall

1991

First published 1991

Copyright © Contributors, 1991

All rights reserved

ISBN 0 901272 85 X

Produced by Chase Production Services
for Lancaster University

Printed and bound in United Kingdom

CONTENTS

NORTH-WEST REGIONAL STUDIES

This volume is the nineteenth in a series of occasional papers in which contributions to the study of the North West are published by the Centre for North West Regional Studies in the University of Lancaster and are available from there. The general editor will be pleased to consider manuscripts of between 10,000 and 25,000 words on topics in the natural or social sciences and humanities which relate to the counties of Lancashire or Cumbria.

ACKNOWLEDGMENTS

The editors would like to acknowledge the kindness and patience of Dr. and Mrs. John Farrer throughout the compilation of this monograph. They are extremely grateful to them for permitting quotation from their private family papers, and for arranging for Mrs. Mary Farnell to make copies of material in their archive.

They are joined in these thanks by Sara Mason, who is also appreciative of help received from Stan Lawrence and Martin Gillibrand. Audrey le Lièvre would like to thank the Regius Keeper of the Royal Botanic Garden, Edinburgh for permission to quote from two letters in the archives.

The guidance received from Drs. G. Halliday and J. Rodwell helped the editors to negotiate nomenclature, although remaining mistakes are theirs and not their advisors'. Support from Dr. Elizabeth Roberts and more help and interest from Mrs. P. Lawley than should be expected from any typist greatly eased a complex task.

The editors gratefully acknowledge the generosity of the following individuals and institutions in providing illustrations: Dr. & Mrs. J. Farrer (front cover, 1, 4, 6, 10, 14, 18, 19, 20, 21, 23, 24, 26, 27); A. D. Schilling (9, 11); The Royal Botanic Garden Edinburgh (13, 15, 16, 17); The Lindley Library of the Royal Horticultural Society (7, 8, 12, 22) and The University of Liverpool Botanic Garden at Ness (3).

PREFACE

This monograph is published by Lancaster University's Centre for North West Regional Studies to coincide with an exhibition prepared by the Peter Scott Gallery as a tribute to the life and work of Reginald Farrer, for display in Lancaster in the summer of 1991, and in the Royal Botanic Garden, Edinburgh in the autumn of 1991. The multi-disciplinary approach represented in this present volume offers the first comprehensive view of the lasting significance of his work.

Visitors have always been welcome to walk along the four mile 'Reginald Farrer Trail', which begins in Clapham and passes alongside the lake and the gorge overlooking some of Reginald Farrer's own plantings. The Ingleborough Estate makes a small charge at the beginning of the Trail, and provides a map. However, it should be noted that Ingleborough Hall itself is now an Outdoor Education Centre, and neither the Hall nor its grounds are open to the public.

The editors have tried to ensure that plants names agree with the rules of nomenclature of the *International Code of Nomenclature of Cultivated Plants* (1980) and the supplement to Bean's *Trees and Shrubs Hardy in the British Isles* (1988). Where the English names of wild flowers are used, these are presented as recommended by the Botanical Society of the British Isles in J.G. Dony, S.L. Jury and F. Perring *English Names of Wild Flowers* (1988), except that on occasion the inclusion of Latin names has enabled us to retain a contributor's preferred common name.

ILLUSTRATIONS

1

AN INTRODUCTORY TRIBUTE TO REGINALD FARRER

William T. Stearn

Rock garden enthusiast and field botanist, writer, traveller, artist, plant collector, Theosophist then Buddhist, to be all these and endowed with exuberant industry and energy, as was Reginald John Farrer (1880–1920), is to stand out anywhere as a many-sided individualist. He stood out thus in the world of Edwardian horticulture, even though gardeners so notable as William Robinson, Maria Theresa Earle, Ellen Ann Willmott, Edward Augustus Bowles, Clarence Elliott, Harry Veitch and Henry John Elwes were his contemporaries. The subtitle 'A book of joy in high places' to Farrer's *Among the Hills* reveals the mainspring of his activities. From a boyhood in Yorkshire near Ingleborough Fell, once reputed to be England's highest mountain, to a tragic death at 40 among the mountains of Upper Burma, he was 'never so happy as wandering in wild places', truly 'a creature of the hills', as his friend E.H.M. Cox described him in 1930. He early became interested in mountain plants when roaming as a boy alone over the Ingleborough hills. From them he brought home plants to identify with the aid of Bentham and Hooker's *British Flora*, of which a fifth edition had appeared in 1887. A notable Ingleborough find was a little mountain member of the Caryophyllaceae now known as *Arenaria norvegica* subsp. *anglica*, a Yorkshire endemic, which had been recorded for the first time, as *A. gothica*, in 1889 and for which Farrer, aged 14, recorded a second station in 1894 (*J. Bot. (London)*, 32, p.344). Some years later he found on Ingleborough an unrecorded hybrid saxifrage (*Saxifraga hypnoides x S. tridactylites*) which Druce named *S. x farreri*, the first and least distinguished of the many plants named in his honour. An article, 'Some British alpines', in *Flora and Sylva*, 3 (1905), pp. 330–334 records his firsthand experience of Ingleborough plants in the wild and in his garden.

Thus by 1899, when Farrer became a student at Balliol College, Oxford, he already possessed a firsthand knowledge of the British montane flora and had begun to grow alpine plants on the limestone rock garden he had made at Clapham, his home. He found in Oxford a like enthusiast for growing alpine plants, the Rev.

1. Reginald Farrer wearing robes collected on his travels.

Henry Jardine Bidder (1847–1923), the Bursar of St John's College, and he helped Bidder to make the extensive St John's rock garden. Unfortunately he had been born with a hare-lip, a disability which he disguised by a thick moustache, but he could not likewise disguise an associated high-pitched voice. His friends ignored this. Farrer, however, was sensitive about it; he has indeed been described as that rare and strange phenomenon 'a Yorkshireman with an inferiority complex'. It did not prevent him from making many lasting friendships among horticulturists such as E.A. Bowles and Clarence Elliott. A more surprising friend was Harold Brewer Hartley (1878–1972), a student of chemistry and mineralogy, who entered Balliol College in 1897 and became surreptitiously involved in creating large rubies artificially; later he was in charge of the Chemical Warfare Department with the rank of Brigadier General in the 1914–1918 World War but he had wide interests and Farrer gave him many landscape paintings.

After graduating in 1902, Farrer set off on the first of his long journeys, one to Japan and Korea with a visit to Peking (Beijing). Japan entranced him, particularly Tokyo as it was then, 'this flowery stinking adorable city'. Visiting its gardens he learned more quickly than most Europeans that 'the object of a Japanese garden is not to be a paradise of flowers but a reproduction of a landscape' made restful by its greenery and carefully chosen aesthetically placed rocks. Flowers did not intrude into such Japanese gardens but were grown outside and subject to certain rigid rules of taste. 'At the head of rejected blossoms', Farrer noted, 'stand the rose and the lily, both of which are considered by the Japanese rather crude efforts of nature. Many others, of less beauty, fall under this condemnation. The elect are cherry, wisteria, peony, willow-flower, iris, magnolia, lotus, peach, plum and morning-glory'. This oriental stay resulted in a book of 296 pages, *The Garden of Asia. Impressions from Japan* (1904).

By now Farrer had begun his career as a prolific author, unsuccessfully writing novels, successfully describing rock garden plants and their cultivation. He published in 1906 *The House of Shadows*, in 1907 *The Sundered Streams* and *My Rock-Garden*, the most frequently reprinted of all his books. There followed in 1908 *Alpines and Bog-Plants*, *The Ways of Rebellion*, *The Dowager of Jerusalem* and, as the result of a journey, essentially a Buddhist pilgrimage, to Ceylon (Sri Lanka) in 1907, *In Old Ceylon*. Nevertheless writing did not monopolise his time. He visited gardens in Britain and Ireland, including Miss Willmott's celebrated long rock garden at Warley resembling a mountain gorge, then at its floriferous heyday, now over-shadowed by sycamores and without a single rock plant. He attended meetings of the Royal Horticultural Society in London. Above all he made excursions to the Alps, once with Clarence Elliott (1881–1969), four times with E.A. Bowles (1865–1954). Elliott, who spent an exhausting month in the Alps with Farrer, found him 'a tireless traveller, a great walker and a fearless climber', 'a curious, complex fascinating personality'.

Farrer published two more books in 1909, *In a Yorkshire Garden* and *The Anne-Queen's Chronicle*, the most readable of his mostly unreadable novels, then in 1911 *Among the Hills*, in 1912, *Through the Ivory Gate* and *The Rock-Garden*, in 1913 *The Dolomites, Laurin's Garden* and *Vasanta the Beautiful*. To have achieved all these by the age of 33 was indeed a remarkable performance. Clarence Elliott rightly stated in 1921 that 'As a writer of garden books he stood alone. He wrote, as a rule, from a peculiar angle of his own, giving queer human attributes to his plants, which somehow exactly described them. He wrote vividly, often at the top of his voice as it were. . . . The chief reason why his books are so valuable is that one can read them. He always had something to say'. He obviously enjoyed writing them.

Some time before 1913 Farrer had embarked on an immense ambitious undertaking, an encyclopaedia of rock garden plants including not only those in cultivation or which he knew personally in the wild but also others which, from descriptions in botanical literature, appeared worthy of introduction and cultivation. He completed this, *The English Rock-Garden*, in 1913 but the 1914–1918 World War delayed publication until October 1919. Here he indulged to the full his taste for vivid indeed sometimes extravagant writing and expressed his strong likes and dislikes. Accordingly it received a mixed reception. 'The first thought on dipping into this book at random', wrote a reviewer in the *Gardeners' Chronicle* of December 1914, 'is what an invaluable book it might have been if only Mr Farrer could have ceased for a while to be Mr Farrer. . . . Then we should not have had our pleasure dimmed by extravagant and fantastic language, irrelevancies, perversities, and three words where one would do'. A reviewer in the *Journal of Botany* (December 1918) remarked that 'we have seldom met with a work wherein the author's self-satisfaction was so conspicuous'. The result, however, of this derided verbal exuberance is that nowadays one tends to consult these two massive volumes primarily for entertainment, not so much to learn about individual plants as to relish how picturesquely Farrer describes them.

His accounts of some genera are certainly alluring. Thus, though admitting that members of the genus *Epimedium* are 'much of a muchness', he then described them as 'plants of extreme but unappreciated value for quiet shady corners of the rock garden, where they will form wide masses in time, and send up in spring and early summer 10 inch showers, most graceful and lovely, of flowers that suggest a flight of wee and monstrous columbines of waxy texture, and in any colour, from white, through gold, to rose and violet. Then beginning later than these, appear the leaves, hardly less beautiful an adornment to summer than the blossoms to spring. For these are of a delicious green, much divided into pointed leaflets, and borne on airy wiry stalks'. Until I read this enthusiastic description I had never even come across the name *Epimedium*; thereafter, my curiosity aroused, I sought epimediums in gardens and I found them indeed as attractive as Farrer indicated. Ten years later, in 1938, after studying herbarium specimens in 38 botanical institutions and living plants obtained from 22 gardens, with the 'dire tangle' (Farrer's challenging phrase) of taxonomy and nomenclature at length unravelled, I published a monograph of *Epimedium* and *Vancouveria* covering them all, but Farrer had in a sense initiated this fascinating self-imposed task.

Here and there, notably in dealing with *Primula megaseifolia* and *P. marginata*, Farrer's malicious humour surfaces. Thus of *Primula marginata* he wrote: 'Naturally it varies copiously, and the gardener had best go and choose his forms. He is particularly recommended to go to the valley of La Maddalena, above San Dalmazzo de Tenda, not only because there *P. marginata* exists in the most rampant profusion and the most riotous and lovely degree of variation, but because the valley is further occupied by a famous English botanist, one Mr Bicknell, who has there a house and spends long summers, in the course of which he asks nothing better than to show the treasures of his hills to all such fellow-collectors as desire to see them. Therefore, in asking him for guidance, the gardener will not only be gaining profit but giving pleasure also – a holy and pleasing thought.' Having received a frosty reception at La Maddalena, Farrer was well aware how much the Rev. Clarence Bicknell (1842–1918), author of *Flowering Plants and Ferns of the Riviera* (1885), hated visitors to his valley! Farrer's barbed remarks, however, missed their target. Publication of *The English Rock-Garden* was delayed until October 1919; Bicknell died in July 1918.

Farrer found Bowles an especially valuable friend not simply for his horticultural knowledge, experience and enthusiasm but also for his wise, tolerant, steady and reliable character and genial sense of humour. Farrer's dedication of *Among the Hills* (1911), 'Ave Crocorum omnium Rex Imperator Paterculus Angustus' [Hail King Emperor of all Crocuses Little Father Augustus] is to Bowles. He admired both Bowles and his unpretentious rock garden overflowing with a diversity of well-grown plants at Myddleton House, Enfield. His preface to Bowles's *My Garden in Spring* (1914) brought, however, much embarrassment to Bowles through a stupid misunderstanding by Sir Frank Crisp.

Publications by the Russian botanist Carl Johann Maximowicz (Maksimovich) on botanical collections made in Kansu (Gansu) and adjacent Tibet by the Russian explorers Potanin and Przewalski indicated that this seemingly bleak north-western region of China possessed a flora rich by European standards even if less so than Yunnan and Szechwan (Sichuan) but much more likely to yield good garden plants of sure hardiness in Britain. Work on *The English Rock-Garden* probably brought these publications and the region to Farrer's attention. Deciding to visit Kansu he had the supreme good fortune of getting the services as companion and assistant of William Purdom (1880–1921), a Kew-trained gardener from Westmorland, who had already collected in China, including Kansu, from 1909 to 1912 for Messrs Veitch of Chelsea and the Arnold Arboretum. He was a steady, even-tempered, resourceful and practical man, through whom alone, as Farrer said, 'these odysseys were made possible and pleasant'. They travelled across Siberia by the Trans-Siberian Railway to Peking (Beijing), Purdom going ahead of Farrer to make travel arrangements. 'So', to quote Alice Coats, 'the incongruous pair set out – Farrer harsh-voiced, excitable, heavily moustached and tending to fat; Purdom tall and lean, of magnificent Nordic physique, equable and reticent, but to Farrer "an absolutely perfect friend and helper"'. On 5 March 1914 they travelled westward from Peking and reached Kansu on April 8. The most exciting of their immediate finds was the *Viburnum* which Bunge had described in 1833 as *V. fragrans* from plants cultivated in Peking but for which the nomenclaturally correct name is now *V. farreri*. In 1835 Potanin had already found it in a wild state in Kansu and later Purdom had introduced it into cultivation, but the plants which Farrer saw on 16 April 1914 he enthusiastically though erroneously claimed to be the first discovered in a genuinely indigenous state. Their two years (1914 and 1915) of collecting specimens and seeds in Kansu, despite the lawlessness of the country, were described at length by Farrer in his *On the Eaves of the World* (2 vols., 1917) and his posthumous *The Rainbow Bridge* (1921).

Farrer returned to London for war work. Purdom remained in China, becoming forestry advisor to the Chinese government. His sister told me that he aimed to restore lost woodland to the Chinese and help them to preserve their forest resources, but he died in Peking in 1921, a year after Farrer's death. Plant names with the epithet purdomii in the genera *Aster, Astragalus, Buddleja, Caraganas, Coluria, Delphinium, Dracocephalum, Leptodermis, Ligularia, Populus, Primula, Rhododendron, Sedum* and *Thalictrum* commemorate him. Farrer had obviously established good relations with the Royal Botanic Garden, Edinburgh; to its herbarium already rich in Chinese specimens collected by George Forrest went his. Years of study of Forrest's material by the Edinburgh Professor of Botany and Regius Keeper, Isaac Bayley Balfour (1853–1922), the Deputy Keeper, William Wright Smith (1875–1956), and the Keeper of the Herbarium, John Frederick Jeffrey (1866–1943), had given them experience unrivalled elsewhere in the identification of Chinese plants. Maximowicz in St Petersburg, Franchet in Paris and Hemsley at Kew had been solitary workers on the Sino-Himalayan flora. At Edinburgh its study had become an institutional speciality. Farrer's material showed the Edinburgh staff that his Russian predecessors had by no means gathered all the plants of Kansu; the *Notes from the Royal Botanic Garden, Edinburgh* accordingly contain the descriptions of many new species based on Farrer and Purdom's specimens. Their aim, however, was not to increase botanical knowledge but to add worthy new hardy plants to gardens. Although their seeds reached Britain at a sad and most unfortunate time, with garden staffs depleted by the 1914–1918 World War, nevertheless the presence in gardens of *Allium cyathophorum* var. *farreri, Anemone tomentosa, Aster farreri, Buddleja alternifolia, Deutzia longifolia* var. *farreri, Gentiana farreri, Geranium farreri, Jasminum humile* var. *farreri, Lilium duchartrei,*

Meconopsis quintuplinervia, Paeonia veitchii var. *woodwardii, Potentilla parvifolia, Rodgersia aesculifolia, Rosa elegantula* var. *persetosa, Semiaquilegia ecalcarata, Syringa potaninii* and *Viburnum farreri* testify to the achievement of Farrer and Purdom's aim.

Before the triumphant publication of *The English Rock-Garden*, Farrer had set out for Upper Burma with Euan M.H. Cox (1893–1977) as his companion. Their chosen region, the country around Hpimaw, about 150 miles north-north east of Myitkyina, had already been visited by Forrest and Kingdon Ward. Cox returned to England from Rangoon early in 1920, but Farrer returned to Upper Burma, where he died in the Ahkyang Valley on 16 October 1920. This expedition likewise yielded species new to science, including *Notholirion campanulatum, Nomocharis farreri, Omphalogramma farreri* and *Picea farreri*, but was not horticulturally so successful as that to Kansu owing to the climatic differences between Upper Burma and the British Isles, as noted by E.H.M. Cox. Farrer contributed to the *Gardeners' Chronicle* 39 articles between June 1919 and March 1920 describing the expedition, but Cox's book *Farrer's Last Journey, Upper Burma 1919–20* (1926) provides the final account.

Looking back over Farrer's writings, his contributions to periodicals (of which my list in Cox's *The Plant Introductions of Reginald Farrer* (1930) is extensive but not complete) as well as his 19 books, one cannot but be impressed that he had accomplished all this and much else within 20 years. He was a complex highly talented and many-sided individual but his skill as a painter of plants and landscapes, for which, much to his regret, he had no training whatever, came as a surprise when in 1918 the Fine Art Society staged an exhibition of 201 water-colours painted in Kansu and eastern Tibet.

Cox said of Farrer that 'he was one of the few men I have met of his mental calibre who could live entirely on themselves'. The key to this was, I think, his literary urge. Kingdon Ward had a like temperament for enduring solitude in the wilds. A prolific writer, Kingdon Ward likewise sought continuously to embody in apt words the multitudinous impressions with which plants, people and places fed his restless soul. Writing is essentially talking to oneself, of enjoying one's own solitary company. It gave both men an inner strength for achievement.

A sketch of Farrer's career as outlined above can provide no more than the background to appreciation of his activities, which others have treated in more detail in the present commemorative volume. From that stands out his uniqueness. No one will ever again enrich and enliven the literature of horticulture as Reginald Farrer did.

2

PLANT COLLECTING AND IMPERIALISM

John M. MacKenzie

It has often been said that one of the first things the British did at any new imperial settlement was lay out a race course. If that is true the second to be laid out was a garden. In reality, the garden often came first, for in most parts of the world botany and the British were inseparable. In the nineteenth and early twentieth centuries an interest in botany was the mark of all cultivated Britons overseas. They filled their letters, journals and memoirs with botanical information; they drew and pressed flowers and other plants; and they regarded the ubiquitous botanical garden as an important sight on their empire tours.

There are a number of reasons for this fascination. At one level, the British were merely extending overseas an important upper and middle class interest from home; but a global botanical enthusiasm ran much more deeply than this. It had economic, medical, taxonomic, and sentimental dimensions. It involved global redistributions of plants, improving on nature by seeking to find the ecologies to which species from other continents could be transferred. The search for garden plants to enhance the range, diversity and aesthetic attractions of gardens at home was but one facet of an immensely diverse activity.

From the seventeenth century, the colonies of settlement in the New World sought to discover cash crops which would constitute the export staple upon which the colony's economy could be built. As is well known, among those that were found to grow well in the southern part of North America and the West Indies were cotton, tobacco and sugar cane. These were labour intensive crops which helped promote the slave trade to satisfy the labour requirements of plantations. This introduced another botanical imperative: to discover suitable crops which would feed slaves in the most secure and economical way. It was for just such a purpose that the ill-starred Captain Bligh set out on *HMS Bounty* in 1787 to collect specimens of the bread-fruit tree in the South Seas. Elsewhere, gardens were founded to grow fresh fruit and vegetables for mariners in order to avoid the problem of scurvy. Typical of such gardens were those at St Helena and the Cape of Good Hope.

The second half of the eighteenth century was the era of the establishment of botanical gardens, matching the foundation of Kew at home. Famous gardens were created in the West Indies (St Vincent in 1763), Mauritius (the celebrated Pamplemousses garden founded in 1769 by Pierre Poivre during the French occupation of the island and later taken over by the British), and by East India Company officials in Madras in 1778 and Calcutta in 1786. In the nineteenth century such 'gardens of acclimatisation' became the norm. Famous gardens were founded at Trinidad (1820) and in the 1840s and 50s at Darjeeling in India, Peradeniya in Ceylon, Hong Kong and Cape Town (among many others) to experiment with suitable plants. Mission stations in Africa invariably laid out gardens for the same reason.

Some of these gardens, linked through Kew in an extensive imperial research network, helped to create the mushroom growth of tea planting in northern India and Ceylon, cinchona planting in the Nilgiri Hills of south India, rubber in Malaya (first acclimatised at the Singapore garden in 1891), cotton in the Middle East and spices on various oceanic islands. Later came coffee, cocoa (in this case introduced from the Caribbean by an African), tea, tobacco, cotton, and sisal in Africa. Sugar cane was also grown on Indian Ocean islands, in Queensland and Natal. These are but a few of the most notable examples of imperial crop transfers which helped to shape the economies of the modern world.

These activities have been portrayed as a global conspiracy to remove plants from their original environments, create competing sources of production and thereby damage the economic health of the areas in which the plants were native species (this has been argued, for example, in relation to cinchona and the Andes, rubber and the Amazon, tea and China).[1] There can be no doubt that indigenous peoples often resisted the plant hunters, suspecting them of detracting from the natural riches of their areas by removing specimens. Reacting to these suspicions, plant hunters sometimes attempted to work clandestinely or adopted some form of disguise, contributing to the sense of secrecy and adventure associated with their trade.[2] It is, however, hard to see how plants could ever have been protected within specific environments once modern communications systems and scientific studies had been established. 'Plant protectionism' was never a practical policy.

Botany has of course always been associated with medicine. Many of the early botanists were doctors, seeking plants with curative properties. This was no less true of the British Empire. In India, the earliest European botanists were invariably East India Company surgeons. Discovering prophylactics and cures for tropical diseases was an important aid to European settlement in the tropics. Malaria was a major scourge in Africa and the East and the extraction of quinine from the bark of the cinchona tree from South America was one of the important medical advances of the nineteenth century, well known long before the cause of malaria was discovered at the end of the century. Growing cinchona in India and elsewhere was a response to the need to have extensive and secure supplies. There were many other less prominent examples, including Joseph Rock's search in Burma of the 1920s for the tree which was said to have curative properties in relation to leprosy.

Not all of this botanical activity had strictly utilitarian objectives. Plant hunters were driven by an insatiable desire to collect as many species and varieties as possible in pursuit of the taxonomic ordering of the globe. In this respect, science and the ideology of empire marched together. Imperialism in the nineteenth century was concerned with the re-ordering of the world in the image of Europe. It aimed to create a global economy with continental specialisations linked to industrialism, controlled in effect through the export of the European moral order as epitomised by the phrase Christianity, commerce and civilisation. Science both influenced and contributed to these processes by seeking to classify and categorise the known natural phenomena of the globe. Geology, zoology and botany were all prominent in this respect.[3] Among the greatest achievements of the century were the discovery and application of geological stratigraphies and the extension of Linnaean classificatory systems to the whole range of biological and botanical discoveries.

It would be wrong, however, to think solely in centripetal terms of botanists and plant hunters searching the globe for materials to bring back to the museums, herbariums and gardens of Europe. Many of the plants were intended, after propagation and study, for re-export to new ecologies where they had not previously flourished. There was also an extensive export of British and European plants to gardens overseas. It is at this point that we encounter what might be called the 'sentimental' aspects of plant transfer. Settlers in North America, Australasia and southern Africa wished to be reminded of home. They created gardens with familiar plants, often mixing promiscuously with more exotic local varieties. For the same reasons they acclimatised song birds, deer and many other animals from home, sometimes (as in the case of the rabbit in Australia) with disastrous consequences.[4] The garden plants, of course, invariably 'escaped' and began to colonise their new homes by growing in the wild.

Reginald Farrer and intrepid plant hunters like him should therefore be placed in a much wider context of botanists, doctors, plantation owners, scientists at government field stations, missionary gardeners, seedsmen, and local enthusiasts who were bound up in these dramatic and rapid processes of botanical exchange. The plant hunters themselves came in a variety of different guises, employees of Kew, botanists attached to geographical expeditions and sometimes even military campaigns, naval officers and surgeons, colonial officials, wealthy travellers, and private enterprise collectors funded by seedsmen or by the owners of great gardens in Britain. By the twentieth century we can add university academics to this list. Among the Kew researchers were Francis Masson (1741–1805), who went to the Cape (still under Dutch control) in 1772, Richard Spruce (1817–92), and Sir Joseph Hooker (1817–1911), who subsequently succeeded his father Sir William as the director of Kew. David Douglas (1799–1834) and Robert Fortune (1813–80) both collected for the Royal Horticultural Society, Fortune later collecting tea plants for the East India Company. Among expedition botanists were the great Sir Joseph Banks (1743–1820), who accompanied Captain Cook on his first voyage (1768–71), and Sir John Kirk (1832–1922), who acted as botanist to David Livingstone's Zambezi Expedition of 1858–63. The Government of Bombay appointed a botanist to accompany Napier's military campaign in Abyssinia in 1867. George Forrest (1873–1932), E.H. Wilson (1876–1930) and Frank Kingdon-Ward (1885–1958) all worked for seed companies and wealthy garden owners.[5]

These botanist and collectors often operated at and beyond the frontiers of formal and informal empire and many of them endured real privations in relation to climate, disease, and the hostility of local peoples. But they were nonetheless dependent upon the powerful presence of a great imperial power. They reached their destinations by imperial shipping routes, naval vessels and railway lines. Shipping lines provided a reasonably reliable means of freighting their specimens home (although in the early days there are several examples of much work lost through shipwreck).[6] Many were funded by official and semi-official bodies concerned with scientific advance and its economic application. Inevitably, many were deeply implicated in the ideologies and objectives of imperialism. Moreover, they were often polymaths. They frequently hunted to survive and took a considerable interest in fauna as well as flora, and collected ethnographic, geographic and geological information. Sometimes they were the first to penetrate particular areas, particularly the remoter regions of the Himalayas and the border territories of Assam, Burma, and China, and the information they provided reached far beyond the purely botanical. Joseph Hooker's map of Sikkim, for example, compiled in 1848–50, remained in use for many years until there was a proper survey in the 1880s.[7]

This rich diversity of activity within an imperial framework is perfectly illustrated by the work of J. Forbes Royle and Joseph Hooker in India. Royle was the superintendent of the Honourable East India Company garden at Saharanpur. A doctor, he was also in medical charge of the Company's station, superintending two hospitals and the health of the European residents. As if that were not enough, he was an energetic polymath who took meteorological and astronomical observations and collected plants (which were drawn by the Company's plant painters), specimens of rocks, insects, and the skins and skeletons of mammalia, birds, and reptiles. In 1839 he published magnificent illustrated volumes on 'the botany and other branches of the natural history of the Himalayan mountains and of the flora of Cashmere', which he dedicated to the directors of the East India Company. His plan of the garden at Saharanpur (plate 2) neatly illustrates the many functions of such a garden. It contained Linnean, medicinal and agricultural gardens, a horticultural department, nurseries, conservatories and, in addition to all the necessary services, a Hindu temple and Suttee monuments.[8]

Joseph Hooker was persuaded to go to India (rather than the Andes) in 1847 by
Dr Falconer the superintendant of the East India Company garden at Calcutta, who
pointed out to him that the eastern part of the Himalayas was shrouded in a 'mystery
equally attractive to the traveller and the naturalist'. Hooker was fortunate enough
to travel by the Admiralty vessel *HMS Sidon* which was conveying the Marquis of
Dalhousie, the Governor-General, to India. He was invited to join Dalhousie's suite
and travelled on from Egypt in the *Moozuffer*, a steam frigate of the Indian Navy.
Lord Auckland suggested he might go to Sikkim, 'whose ruler was all but a
dependant of the British Government'. In fact Sikkim was hotly disputed as a sphere
of influence between the Company and China and Hooker found himself embroiled
in what was effectively a border dispute; he was seized by what he described as a
faction of the Sikkim court which preferred allegiance to China. Hooker escaped
and succeeded in collecting many plants. His cool description of these events

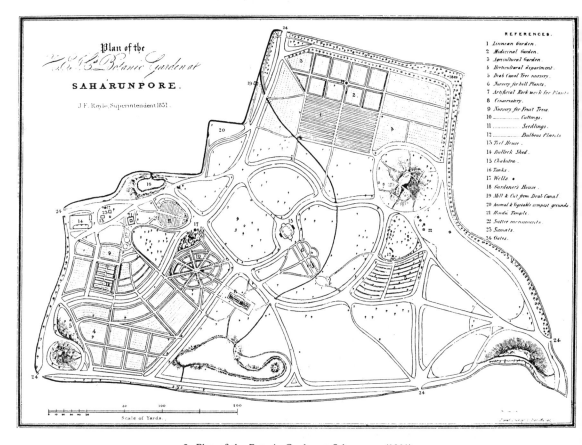

2. Plan of the Botanic Garden at Saharanpur (1831)

abounds in observations not only on natural history but on matters economic, political, social and maritime relating to British India.[9]

Later in the century, the focus of the colonial gardeners was changing; there was a definite shift from utilitarian to ornamental and scientific functions in colonial botanic gardens. Many more were founded in the last decade of the century and the early years of the twentieth. Governors in colonial territories invariably planned fine gardens around their Government Houses. Sir Harry Johnston in Nyasaland (now Malawi) and Sir Hesketh Bell in Uganda both sought to create gardens in relatively primitive circumstances and described their activities in their memoirs.[10] Guide books of empire were full of descriptions of botanic gardens that should not be missed. Sir Algernon Aspinall in his great guide to the West Indies listed no fewer than a dozen.[11] Murray's *Handbook for Travellers in India, Burma and Ceylon* included many in the sub-continent and ecstatically described the walks and plants that could be seen at Peradeniya.[12] The Union Castle Line's *South and East African Year Book and Guide* described the Cape Town garden as 'one of the greatest attractions' of the Cape, while of Pamplemousses it announced 'No description of these gardens can do justice to their beauty'.[13] Charles Kingsley wrote a glowing description of the botanic gardens at Trinidad in *At Last*.

When we enjoy the great array of garden plants, shrubs and trees brought back to this country by the plant hunters we should not forget that this was part of a complex process of exchange distributing plants among all the continents. Botanical exploits were but one aspect of wider economic and scientific processes which saw the fauna as well as the flora of the world dispersed from its natural habitats. We should also remember that the British took their passion for gardens into many other parts of the world and that gardens great and humble, mixing plants from several continents, are to be found in most of the territories which formerly made up the British Empire.

REFERENCES

1   Lucile H. Brockway, *Science and Colonial Expansion: the role of the British Royal Botanic Gardens* (Academic Press, New York, 1979).

2   Robert Fortune, *A Journey to the Tea Countries of China; including Sung-Lo and the Bohea Hills; with a Short Notice of the East India Company's Tea Plantations in the Himalaya Mountains* (John Murray, London, 1852); Hinson Allan Antrobus, *A History of the Assam Company, 1839–1953* (Constable, Edinburgh, 1957) provides an account of the activities of the botanists Dr N. Wallich, superintendant of the Botanical Garden at Calcutta, and Dr H. Falconer, superintendant of the Botanical Garden at Saharanpur, in securing and propagating tea plants from China for Assam and the Nilgiri Hills.

3   Robert A. Stafford, *Scientist of Empire: Sir Roderick Murchison, scientific exploration and Victorian Imperialism* (Cambridge University Press, 1989); see also John M. MacKenzie (ed.), *Imperialism and the Natural World* (Manchester University Press, 1990).

4   Alfred W. Crosby, *Ecological Imperialism: the biological expansion of Europe, 900–1900* (Cambridge University Press, 1986).

5 Charles Lyte, *The Plant Hunters* (Orbis, London, 1983); Charles Lyte, *Frank Kingdon-Ward: the last of the great plant hunters* (John Murray, London, 1989); E.H. Wilson, *A Naturalist in Western China* (Methuen, London, 1913); Fortune, *Journey to the Tea Countries*.

6 See, for example, Alfred Russell Wallace, *My Life a Record of Events and Opinions* (Chapman, London, 1905) and Lyte, *Plant Hunters*.

7 Mea Allan, *The Hookers of Kew, 1785–1911* (Michael Joseph, London, 1967).

8 J. Forbes Royle, *Illustrations of the Botany and other branches of the Natural History of the Himalayan Mountains and of the Flora of Cashmere*, (2 vols, William H. Allen, London, 1839).

9 Joseph Dalton Hooker, *Himalayan Journals: Notes of a Naturalist in Bengal, the Sikkim and Nepal Himalayas, the Khasia Mountains etc.* (2 vols, John Murray, London, 1855).

10 Sir Hesketh Bell, *Glimpses of a Governor's Life* (Sampson Low, Marston, London 1946); Sir Harry Johnston, *The Story of My Life* (Chatto and Windus, London, 1923).

11 Algernon E. Aspinall, *The Pocket Guide to the West Indies* (Edward Stanford, London, 1907). This, the first edition, listed nine gardens, but see also the edition of 1931.

12 *A Handbook for Travellers in India, Burma and Ceylon*, 10th edn (John Murray, London, 1919), p.659.

13 *The South and East African Year Book and Guide for 1930* (Sampson Low, Marston, London, 1930), pp.372, 823.

3

PLANT COLLECTING AND PATRONAGE

Brenda J. McLean

Farrer enjoyed many plant-hunting holidays in the European alps, but when, in 1913, his ambitions reached to collecting in China, the problem of funding began. It seems that his father's allowance, which was his main source of income all his life, would not stretch to such an expedition.

Where should Farrer seek financial aid? His mind turned to the Treasury and the Royal Horticultural Society, to wealthy gardeners and nurserymen who might welcome some new and rare plants in return for a share in the expenses of the expedition. But his problem was not new to plant collectors, nor easily solved.

One hundred years before Farrer's aspirations, it needed the powerful arguments of eminent men to persuade the Treasury to finance plant collection. In 1814 Sir Joseph Banks succeeded in obtaining Government funding for Allan Cunningham and James Bowie to collect plants in South America, and later separately at the Cape and in Australia.[1] Even then, financial cutbacks led to Bowie being recalled early. Professor William J. Hooker, at Glasgow University, protested at this 'needless stretch of parsimony'.

Hooker had exceptional energy and drive, but when he became Director of Kew in 1841, he did not immediately obtain the Treasury's authority to employ a full-time collector. His triumph came in 1847 when he pleaded successfully for a collector to go to India.[2] He nominated his son, Joseph D. Hooker, who opened up to the world the horticultural promise of the marvellous rhododendrons growing in Sikkim. However, the last full-time plant collector employed by Kew, Richard Oldham, died in China in 1864 feeling undervalued and underpaid.[3]

When Farrer attempted to raise money through the Treasury, in 1913, he received a strong warning from Professor Bayley Balfour at the Royal Botanic Gardens, Edinburgh: 'Dear Mr Farrer, ... with regard to your proposals for widening the financial basis of your scheme through a grant from the Treasury ... they are bound to fail.'[4] He explained that the Royal Botanic Gardens themselves had been refused comparable grants: 'It is one of the misfortunes under which we labour as compared with our brethren in Germany, for instance, that subventions for such explorations are not part of the recognised vote to our Establishments.'[4]

If a Treasury grant had been available, it is doubtful if Farrer would have been the recipient. Firstly, he was a keen amateur with no botanical training. In correspondence with colleagues, Sir David Prain at Kew pointed out that if a grant were to be given, the Board would appoint the explorer and 'he must be a man who has the faculty for scientific exploration, of which Farrer has given no proof.'[5] Prain was also concerned that a grant to Farrer would lead to 'free supply of seeds for the exclusive benefit of a particular social class'. He also thought it would be unfair to other collectors in China, like Wilson, Kingdon Ward and Forrest: 'To subsidise Farrer's [expedition] when these others were left to private enterprise must to all thinking people appear inequitable.' Although the Royal Botanic Gardens were ready to co-operate with Farrer in other ways, he had to look elsewhere for financial support.

Would the Royal Horticultural Society sponsor Farrer? When Farrer was born, the finances of the R.H.S. were in a perilous state, and it had already foregone its role as a leading sponsor of plant collectors. In the early 1820s, it had taken over this role from Kew, after the death of Sir Joseph Banks, and the withdrawal of Kew collectors.[6] It had dispatched collectors to China, East Africa, South and North America. After a low period, it had then been stimulated in 1842 by the Treaty of Nanking, making the Chinese ports accessible; its Council advised that the 'Horticultural Society immediately avail itself of the opportunity thus offered for the introduction to Great Britain of the useful and ornamental plants of that immense Empire.'[7] In 1843 the Society sponsored Robert Fortune as one of the first English plant collectors in China.[8] However, by the 1860s, the Horticultural Society's financial situation was deteriorating, and already a new circumstance was emerging − the increasing influence of the nurserymen.

The principal nurserymen were sending their own plant collectors worldwide in search of new and rare species. The family firm of James Veitch & Son, begun in 1808, became one of the foremost nurseries sending out collectors and growing and hybridising the new introductions. When Farrer was planning to go to China, Veitch & Sons had already sponsored two collectors there, Edward H. Wilson and William Purdom.

Farrer spread his net to encompass the R.H.S. and a nurseryman. By 1912 the R.H.S. was in better shape, with increased membership, a new hall in Vincent Square and a new garden at Wisley. It was ready to take the initiative again in the introduction of new species. There was a successful Royal International Horticultural Exhibition in London, and the annual report of Council included the statement that 'The appointment of a plant collector will be considered . . . at a later date.'[9] Farrer's application came at an opportune time in the following year.

He was well known in horticultural circles as a leader in alpine gardening. In addition to his books, he wrote articles on alpine plants in the weekly 'Gardeners' Chronicle', and reported on the rock gardens of the 1912 Exhibition, where his

Craven Nursery Co. won the special cup offered by Lady Trevor Lawrence for Alpines. He had himself previously donated to the R.H.S. a Silver Cup for Alpines and a Farrer Cup for Rock Plants. He was a member of the exclusive Horticultural Club which met monthly for dinner and talks at Hotel Windsor, Victoria Street, London, and he gave a lecture there in November 1913. The 'Gardeners' Chronicle' reported: 'In order to have a fair opportunity of widening his experience as a plant collector, he purposed going in the New Year to Western China and Thibet to familiarise himself as far as possible with the Asiatic Alpine Flora in its native habitats.'[10] Sir Frank Crisp proposed a vote of thanks for the lecture and 'expressed the hope that his journey to China might prove successful and enjoyable.'

3. Bulley's garden at Ness probably about 1910, later to become the University of Liverpool Botanic Gardens at Ness

In the 'Gardeners' Chronicle' in February 1914, it was reported that the R.H.S. had taken a share in the expedition to China organised by Reginald Farrer.[11] The possibility of new and rare species for the new rockery at Wisley seems to have been crucial. There was tempered criticism: a member of the trade asked 'if the Council considered it a wise move altogether that they should help to finance the expenses of an expedition run in the interests of a gentleman who was practically a trader in horticulture', (alluding to Farrer's Craven Nursery Co. although it never seems to have been very profitable.) The Chairman replied that 'they had only involved themselves to the amount of £100 a year for two years, and they felt, as they were really a horticultural society, they ought to try to get the novelties which might come home, as soon as anyone else.' As the R.H.S. had a surplus of about £80,000 at the time, £100 a year was not much, but it went some way to helping Farrer, and set a precedent for other collectors.

Meanwhile, Professor Bayley Balfour wrote that Farrer's 'search for an expense and profit-sharer grows more keen.'[12] Farrer approached a fellow enthusiast, Arthur K. Bulley, who founded his own nursery, Bees Ltd, in 1904. Bulley had solely sponsored George Forrest's first two expeditions (1904—7 and 1910—11) and Kingdon Ward's first expedition (1910—11) in his desire for alpine and hardy plants from China. Bulley and Farrer knew each other well and often exhibited at the same shows. Bulley was a wealthy cotton broker in Liverpool and Farrer had high hopes of him as a sponsor. A lot of bargaining went on, but Bulley contributed more than the R.H.S. to both Farrer's expeditions. For example, Bulley agreed to pay £400 for shares in one season's collecting in 1915, and sent £200 in 1920 when Farrer had cash-flow problems.

Farrer was also supported by other keen gardeners like E.A. Bowles and F.C. Stern. Bowles was on many R.H.S. committees and took great pleasure in his garden at Myddelton House. He and Farrer were great friends, and Farrer dedicated his book 'Among the Hills' to him. They had collected together in the European alps for four successive years, and Bowles admired Farrer's ability to spot a good plant. He sponsored Farrer from personal knowledge of his abilities and the love of new plants for his garden. Stern was creating a now famous garden at Highdown, Goring-by-Sea, and he had already stocked his new rock garden with many Chinese plants collected by E.H. Wilson. He had bought these from Veitch & Sons, and he later described how the initiative of Veitch in sponsoring Wilson, 'encouraged keen people to subscribe to further expeditions to these areas which seemed to contain an unending stream of new and beautiful plants.'[13] Stern was well rewarded for his sponsorship, as seed of Farrer's *Viburnum farreri* (F13) thrived in his garden.

Alpine and rhododendron enthusiasts joined in Farrer's syndicates. An early subscriber, Lt Woodward of Arley Castle, died at the front in WWI and two new plants were named after him, *Primula woodwardii*[14] and *Paeonia woodwardii*[15]. After the war, Major Lionel de Rothschild, of the international bankers, took a full share for two years. He was establishing his 2,600 acre estate at Exbury, near Southampton, and was keen to collect and hybridize new introductions. When Farrer was frantic for more money in 1920, Rothschild also secured another contributor, Baron B. Schröder.[16]

With such widespread support, Farrer found his own solution to the plant collector's perennial funding problem. When Kew had been short of funds in the

1840s, arrangements were made with members of the aristocracy, the Duke of Northumberland and the Earl of Derby, to share the expenses of collectors. Seventy years later, the Royal Botanic Gardens were financially more restricted, but Farrer formed an even larger plant-collecting syndicate. Subscribers ranged from the R.H.S. to keen, wealthy gardeners from many walks of life, from cotton broker to landed gentry. The Royal Botanic Gardens gave their official blessing and were pleased to be involved. In return for herbarium specimens and seeds, they used their expertise in the growth and identification of the new introductions. This combination of expert help and a broadly based syndicate of sponsors was subsequently often used by Kingdon Ward and George Forrest.

NOTES

R.B.G. Edinburgh denotes Royal Botanic Garden, Edinburgh.

REFERENCES

1   Ray Desmond, 'Strange and Curious Plants', in *Plant Hunting for Kew*, ed. F. Nigel Hepper (HMSO, London, 1989), p.7–8.
2   Ray Desmond, 'From rhododendrons to tropical herbs (1820–1939)', in *Plant Hunting for Kew*, ed. Hepper, p.13.
3   Ray Desmond, 'From rhododendrons to tropical herbs (1820–1939)', in *Plant Hunting for Kew*, ed. Hepper, p.15.
4   R.B.G. Edinburgh Archives, File Herb 9/4/1. Expedition to Kansu, 1914. Letter from I. Bayley Balfour to Farrer, 28.9.13. [copy] p.5–6.
5   R.B.G. Edinburgh Archives, File Herb 9/4/1. Expedition to Kansu, 1914. Letter from Sir David Prain to Arthur Hill, 2.10.13.
6   Harold R. Fletcher, *The story of the RHS, 1804–1968* (O.U.P., 1969), p.94.
7   Fletcher, *The story of the RHS, 1804–1968* (O.U.P., 1969), p.148.
8   See also the preceding article by MacKenzie.
9   *Gardeners' Chronicle*, (1912), p.93.
10 *Gardeners' Chronicle*, (1913), p.326.
11 *Gardeners' Chronicle*, (1914), p.116.
12 R.B.G. Edinburgh Archives. File Herb 9/4/1. Expedition to Kansu, 1914. Letter from I. Bailey Balfour to Arthur Hill, 9.10.13.
13 Frederick Claude Stern, *A Chalk Garden* (Thomas Nelson, London, 1960), p.4.
14 R.B.G. Edinburgh Archives, Boxfile Wi-Wo of Balfour, I. Bayley. Correspondence: Letter 29.1.16.
15 R.B.G. Edinburgh Archives, Boxfile Wi-Wo of Balfour, I. Bayley. Correspondence: Letter 26.5.17.
16 R.B.G. Edinburgh Archives, File Herb 7/4/2. Expedition to Up. Burma, 1920. Letter from I. Bayley Balfour to Farrer, 26.3.20.

4

TRAVELLING EASTWARD:
FARRER'S JOURNEYS DESCRIBED

Audrey le Liévre

Reginald Farrer, second cousin of Osbert Sitwell, briefly parades across the stage of Sitwell's autobiography *Left Hand, Right Hand!*[1], a colourful if not always an attractive figure. Sitwell knew that in his own case illness had bestowed upon him in his teens unlooked-for freedom from the dulling atmosphere of boarding school, and time to think and feel, an essential stimulus to his development as an artist. By the same token Farrer's early prowess, his knowledge of plants and skill as a gardener (re-designing the Ingleborough alpine garden at the age of 14) derived from solitary wanderings and observation of the flora among the limestone hills of Yorkshire during a childhood which he spent at home, undergoing frequent operations for the cleft palate with which he was born. So there is a real sense in which all Farrer's journeys started from Yorkshire.

These journeys began when Reginald, aged 3, spent 11 months with his parents in the Mediterranean (and perhaps this early influence stayed with him). At the turn of the century he went up to Oxford where at Balliol he found friends whose interests matched his own. With two of these he spent some months in Japan in 1903, making a brief visit to Korea and apparently travelling via Canada. He recorded his impressions in *The Garden of Asia*[2], published in the following year. Japan stayed in his memory, so that later he wrote 'Some day ... I will have a Japanese build me one of his perfect gardens ... and then I will plant it with my own plants and be happy'. Early in 1908 he accompanied his friend Aubrey Herbert on RMS OPHIR as far as Ceylon, where he disembarked to study flowers and Buddhism (he became a Buddhist at about this time). *In Old Ceylon*[3] was published in 1908. Almost yearly he visited Italy, France or Switzerland, usually in the company of gardening friends: Clarence Elliott, Arthur Bartholomew, Clutton Brock and (after they met in 1910) E.A. Bowles[4] and his friends. A number of plants were brought home which Farrer sought to acclimatise at Ingleborough, where he had now laid out a second alpine garden, his 'Craven Nursery' and a moraine garden. He was also becoming known as a gardening writer: his 1910 'cheap little humble trip', which covered the Graian, Cottian and Maritime Alps, being immortalised in the delightful, if rather over-ebullient *Among the Hills*[5] (1911). *The Dolomites*[6] appeared in 1913.

Farrer always insisted on the importance of mountain solitude to him, though his cousin Osbert felt that his real wish was to shine in the world of conversation and wit where there were many who were captivated by his style, soon ceasing to notice the hare lip and the piercing voice. Perhaps these contrasts formed an essential dichotomy in his character. He was a man of strong likes and dislikes, rooted perhaps in the feeling, unjustified as it turned out, that there were an indefinite number of tomorrows in which matters could be put right (if necessary). Some of those who travelled with him, unused to mountains, clumsy and miserable amid the discomforts of plant hunting in icy rain and mist, felt the measure of his scorn. Sitwell remarked that there were little prods and digs for all in his conversation, and certainly there are in his writings.

Euan Cox[7] was right in observing that this period of work in Europe was only a novitiate (a word well-preferred to 'apprenticeship') in the business of plant hunting. Farrer became fired with the idea of exploration in China while researching his rock garden book. He came upon accounts of journeys by the Russian plant hunters Przewalski and Potanin in Kansu and Tibet and resolved to follow in their footsteps. To avoid clashing with other workers in the field Farrer discussed the areas he should cover in severely practical terms with Isaac Bayley Balfour, Director of the Royal Botanic Garden, Edinburgh. He was fortunate to be introduced to William Purdom, who in 1911 had travelled for Veitch and the Arnold Arboretum in the Tsin-Ling, and who proved to be the ideal companion, tactful, good-tempered, adept at getting the best from the Chinese, smoothing the way for the expedition and allowing Farrer to play impressive variations on the role of the Great Lord.

Apparently with some financial support[8], the two left England for Peking to assemble the elements of their expedition: by April 1914 they were ready to move off from Lanchow. With their wonderfully ragamuffin and individualistic staff (and it is hard to resist Mafu, the Go-go, the Good Little Cook and all the others, with their ponies, especially the intransigent Spotted Fat)[9], they quartered the country from their base at the clement and blossomy little town of Siku, and conquered the Sha-tan-yu ('Satanee') Alps and the terrors of 'Thundercrown'. At the end of the season, when seed had been harvested, winter quarters were found at Lanchow, and here they sampled the highly eccentric, cheerful and funny face of Chinese officialdom in its social guise. Plans for the following season were laid amidst the sadness of contemplating the distant desolation of World War I. In April 1915 they set out again, this time for Sining-fu, whence they searched the Ta-Tung ('Da-Tung') Alps from their base at 'Wolvesden House' – surely the most homely and charming of all their many residences.[10] This, too, came to an end with the season, and in December 1915 the expedition disbanded, Purdom to stay in China and

Farrer to travel home via Petrograd (now Leningrad) where he wrestled with herbarium sheets to sort out the complexities of primula. Farrer joined the Ministry of Information for the rest of the war, and it was not until 1919 that he and Euan Cox set out for Upper Burma, the only area not overrun by other plant hunters.

4. W. Purdom from the Farrer family's album of photographs
from Kansu 1915, annotated "Bill in camp with the boys"

Hpimaw was their base, from which, with their devoted Gurkha staff, they worked an area of about 40 miles. In the winter of 1919–20 Cox had to return home, while Farrer stayed on: after spending some time at Maymyo he travelled up the western branch of the Irrawaddy, striking east to explore the frontier range. The monsoon climate, with its prolonged rain and mist, depressed him: 'Regan and Goneril themselves would have pressed their father to stay the night' he wrote to Euan Cox.[11] He missed the friendly atmosphere of Kansu and became bored with the ever-present woody plants. Here, on 17 October 1920, he died after a brief illness, thought to be diphtheria.

And the results of these trips in terms of new plants?[12] Farrer's clear aim was to collect good garden plants, and by the time *The Plant Introductions of Reginald Farrer* ed. E.H.M. Cox[13] appeared in 1930 some 432 plant names had been determined – only a moderate proportion of the total collected, as Farrer so often omitted to take herbarium specimens. Cox gives a list of garden plants which had proved their worth. In *Plant Hunting in China* (1945)[14] he also lists the most important garden survivors from the Burma expeditions.

But Farrer's great importance lies in his writings, with their flamboyant and larger-than-life quality which transfixes his readers. Who could forget the account of his journeys or fail to see in the mind's eye the travellers clambering, slithering and stumbling over the vertical slopes, frozen cold and wet through, falling into rivers and riding over rickety bridges, but always with an eye for the distant patch of colour which might betoken a new plant and always, mercifully, surviving uninjured to sleep in one more bug-infested, rat-ridden, welcoming Chinese inn. Who could forget certain descriptions, like this one for *Primula optata*: 'It was a curious sweetish scent, as of an old apple cupboard haunted by mice' – or this: 'So many of the Pansies have silly faces: the garden ones one often longs to slap, they look so stupid – like kitchen clocks' or, of a lake high in the Alpes Maritimes: 'Such a clear lake: a tiny little basin of water, like a silver shield forgotten up there among the hills'. Was it Farrer's expeditions I was asked to write about, or his books? It seems hardly to matter now, since they are in effect one and the same thing.

It is only too easy for us to think of Farrer's journeys as we know them from his books, an orderly progression ending in Burma. But to Farrer himself they must have seemed utterly different: Burma held no final importance to him, and just before he died, aged only 40, he was envisaging a trip to Nepal or Tibet for 1923, and for the interim, he had made up his mind to marry and sought the help of his old friends Aubrey and Mary Herbert to find a wife.[15] It is for speculation how a Mrs Farrer might have fitted in to her husband's life, he having made it plain that plant hunting would take priority. Would a happily married Farrer have given it all up to settle in England? Would he have returned to search again the European Alps, to keep nearer to home? Or would he have set off for Nepal, rejuvenated in spirit and more than ever ready for the mountains? Or even, would she have travelled with him, a second Mrs Potanin?

But this didn't happen. Instead, he died, leaving half a lifetime unused. And when his work came to be assessed in obituaries, it was clearly as artist rather than as scientist that he was valued. It has to be admitted that, capable though he was of

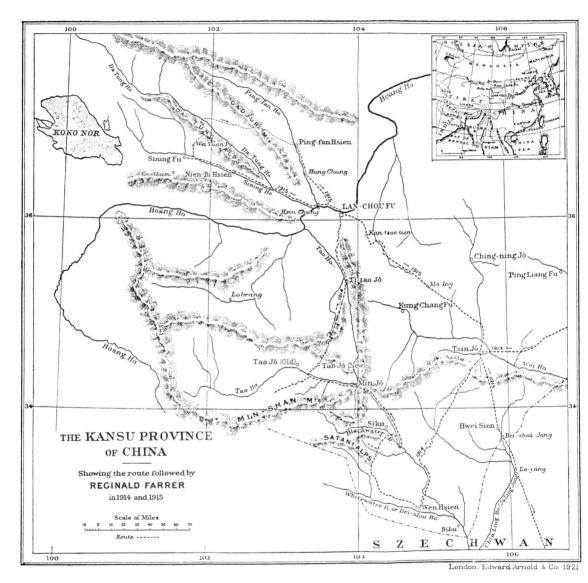

5. Reginald Farrer's route in Kansu, 1914 and 1915

sorting out taxonomical muddles, his approach to the science could be a little cavalier. An impatient comment in a letter to Bayley Balfour makes an apt illustration: 'I'm *sick* of being asked the name of the Alpine Field Campanulas: say what one will, they are as close to rotundifolia as a stamp to an envelope, and more fluctuating in their developments than the heart of woman'.[16] So very un-taxonomical, but so much more memorable. The artist in Farrer, too, has a word to say on *Campanula* (*cenisia*, in this case): '... a clear, pale, rather hard blue, exquisitely brilliant, but softened with a tinge of grey. They look straight up into the eye of day ...' I too think that the artist has the last word.

6. Reginald Farrer's grave

NOTES

Single inverted commas and brackets have been used for nicknames and special versions of Chinese place names used by Farrer.

REFERENCES

1   Osbert Sitwell, *Left Hand, Right Hand!* (5 vols, Macmillan, London, 1945–1950), vol. 2: *The Scarlet Tree*; vol. 5: *Noble Essences*; see also B. Massingham, *A Century of Gardeners* (Faber, London, 1982).

2   Reginald J. Farrer, *The Garden of Asia* (Methuen, London, 1904).

3   Reginald J. Farrer, *In Old Ceylon* (Edward Arnold, London, 1908).

4   Mea Allan, *E.A. Bowles and his Garden at Middleton House, 1865–1954* (Faber, London, 1973).

5   Reginald J. Farrer, *Among the Hills* (Headly Brothers, London, 1911).

6   Reginald J. Farrer, *The Dolomites* (Adam and Charles Black, London, 1913).

7   Euan H.M. Cox (ed.), *The Plant Introductions of Reginald Farrer* (New Flora and Silva, London, 1930), p.8.

8   See also the preceding article by McLean.

9   Reginald J. Farrer, *On the Eaves of the World* (2 vols, Edward Arnold, London, 1917).

10 Reginald J. Farrer, *The Rainbow Bridge* (Edward Arnold, London, 1921).

11 R.B.G. Edinburgh Archives, Letter from Farrer to Cox, 11.9.1920.

12 See also the following articles by Stearn and Schilling.

13 Cox (ed.), *The Plant Introductions of Reginald Farrer*.

14 Euan H.M. Cox, *Plant Hunting in China* (Collins, London, 1945).

15 Margaret Fitzherbert, *The Man who was Greenmantle* (Oxford University Press, 1985).

16 R.B.G. Edinburgh Archives, File Herb 9/4/1. Expedition to Kansu, 1914. Letter from Farrer to I. Bayley Balfour, 14.10.1913.

5

FARRER AND THE VICTORIAN ROCK GARDEN

W. Brent Elliott

Let it be said at the outset: as an historian of the rock garden Farrer was exceedingly unreliable.

> Times have wholly changed for the rock-garden. Fifty years ago it was merely the appanage of the large pleasure ground. In some odd corner, or in some dank, tree-haunted hollow, you rigged up a dump of broken cement blocks, and added bits of stone and fragments of statuary. You called this 'the Rockery,' and proudly led your friends to see it, and planted it all over with Periwinkle to hide the hollows in which your Alpines had promptly died. In other words, you considered only the stones, and not the plants that were to live among them ...
>
> Thanks, however, to the insistent crying in the wilderness of a few ardent evangelists a revolutionary change has passed over the scene in the last twenty years ... now the truth has dawned, and its full daylight is approaching ...[1]

Nowhere does Farrer name these evangelists, but it is apparent that they were English: 'it has been left for England (for the only occasion in recorded history) to head the world in a new and living form of art'.[2]

The passage just quoted vacillates between seeing the Victorian rock garden as the result of a deliberate intention ('you considered only the stones') and seeing it as the result of a technical failure ('your Alpines had promptly died'). As an account of the rock garden in Victorian times, this is a caricature, but it nonetheless hints at the conflict that was raging by the time Farrer was born, between the rockwork as a picturesque construction, and the rock garden as a place for growing alpine plants.

The ornamental rockwork itself had a double origin, as a descendant of the Renaissance grotto and of the naturalistic rock formations of the 18th-century landscape garden. With each succeeding generation there was a new demand for greater naturalism in construction: first, that only natural stone be used (out with corals, shells, and Iceland spar); then, that rocks should be laid on their natural bed (out with collections of rocks arranged as spikes, as can still be seen at Bicton); then, that the rock garden should appear in an appropriate position (out with rockworks emerging from a smooth lawn or adjacent to the parterre). From the 1840s the idea of the geologically accurate construction was particularly associated with the firm of James Pulham and Son, whose artificial stone rockworks, at sites ranging from Battersea Park to Waddesdon Manor and Madresfield Court, set standards of accuracy in the representation of stratification.[3]

Such rockworks were planted in a picturesque manner, with conifers and creeping plants; the plants we now group together as alpines tended to be cultivated in pots, although by the 1830s experiments in cultivating them on rockworks were being made. The justification for such attempts lay in the writings of John Claudius Loudon, whose principle that the garden should be seen to be a work of art, not of nature, led him to recommend planting rocks with exotic alpine species, in order that the rock garden could not be mistaken for nature's work.[4] In the late 1850s, the firm of James Backhouse and Son of York became the first firm to issue catalogues devoted solely to alpine plants.

Backhouse and Son constructed in their nursery a rock garden which was to become influential for its very refusal of naturalistic imitation. 'Rockwork', said C.P. Peach in the 1870s, 'does not want to be an attempt to deceive, but a picturesque place to grow Ferns and Alpines'; geological accuracy was irrelevant to the culture of alpines.[5] The idea was gradually taking root of the alpine rock garden, in which the plants were the important thing, and the stones served only to shelter them.

Standing at the junction of these two traditions was William Robinson's book *Alpine Flowers for English Gardens* (1870). On the one hand, it was primarily a manual on the culture of alpine plants, but on the other it recommended geological accuracy, praised some of Pulham's rockworks (James Pulham was to recommend Robinson's book as a planting guide for his firm's rockworks), and included instructions from Backhouse on construction.[6] By the 1890s, the Backhouse firm, with Richard Potter as its rock garden designer, was creating rock gardens around the country – most notably for Ellen Willmott at Warley Place and, in part, for Sir Frank Crisp at Friar Park.[7]

Farrer's attitude towards Robinson was ambiguous. Much of such history of gardening as can be found in his writings was derived from the already unreliable account in Robinson's various books, and he tended to take at its face value Robinson's image as an iconoclast in the Victorian world: 'suddenly, flaming and audacious, arose Mr Robinson with a crash among the Lobelias of the late Victorian era'. But the apparent admiration for Robinson's alleged revolutionary impact was quickly qualified: 'Like all true prophets, he arose magnificent, passionate, unguided and unguidable. It is the hard fate of the Moses of one generation that he always becomes the venerated rear-guard of the next.' Robinson's actual recommendations Farrer dismissed as a 'naturalism ... purely anarchic, unruddered, unfounded on any rule or depth of knowledge'.[8]

Farrer, indeed, claimed that 'between 1870 and 1890' there was 'an almost general

7. 'Massive rocks (mill stone grit), Friar Park. June 1910'
from Robinson *Alpine Flowers for Gardens* 4th ed., 1910

slump in the noble and exquisite art of cultivating Alpine plants' (though he never specified these great achievements of the pre-1870 period).[9] The two decades thus repudiated were the very period when Farrer would have us believe Robinson's influence was dominant; and in practical terms, it was a period dominated by the competing firms of Pulham and Backhouse. Although Farrer never mentions Backhouse by name, that firm was associated with Robinson; when Farrer said that 'Mr Robinson's wood-cuts of rockwork seem nowadays, to happier knowledge, almost too bad to be even comic', it was the illustrations accompanying James Backhouse's instructions on construction that he meant.[10] When, in *The Rock-Garden*, Farrer illustrated the 'almond-pudding system' of placing rocks, he could

8. 'View of rockwork (a portion) at the York nurseries'
from James Backhouse & Son Catalogue, 1898

have reproduced instead the engraving of the Backhouse rockery which the firm used in their catalogues of the 1890s.[11]

This repudiation of Robinson lies at the heart of much of Farrer's rhetoric, for it was only in his article on rock gardens for *The Horticultural Record*, the book of the Royal International Horticultural Exhibition of 1912, that he spoke openly about Robinson in this way. More often he attacked Robinson obliquely, by attacking someone else for opinions with which Robinson was identified. Even in the *Horticultural Record* article, he went on to associate Robinson's rhetoric of nature with Ruskin.[12] In the preface to his *English Rock Garden*, he further attacked Ruskin for the invention of spurious vernacular names; but the horticultural writer most readily associated with this habit was Robinson, the author of *The Virgin's Bower*.[13] And in his most celebrated diatribe, in his preface to E.A. Bowles's *My Garden in Spring*, he hit at Robinson again, through the person of Sir Frank Crisp.

The story is reasonably well known of how Sir Frank Crisp, sensing in Farrer's preface an insult to his rock garden at Friar Park, took umbrage and, not mentioning Farrer's name, intemperately attacked the innocent Bowles for publishing the passage; and how Ellen Willmott stood at the gates of the 1914 Chelsea Flower Show handing out copies of Crisp's offprint.[14] Willmott's reason for supporting Crisp has usually been attributed to personal motives – frustrated advances on Bowles, for example. Rather more to the point may be the facts that her rock garden had been designed by Richard Potter of the Backhouse nursery, who had also been the designer of part of Crisp's rock garden; and that her rock garden also was characterised by massed planting, as can be seen from Correvon's testimony and the photographs in her own book *Warley Garden in Spring and Summer*.[15]

The further link between Crisp and Willmott is Robinson, who accepted Crisp's piece for publication in his magazine *Gardening Illustrated*, and who had described the rock garden at Friar Park as 'the best natural stone rock garden I have ever seen', a quotation Crisp used prominently in his article.[16] And Robinson was the most public advocate of the principle Farrer was most concerned to attack: the massing of colour in the rock garden. First, here is the controversial passage from Farrer's preface:

> ... carpet-bedding is bursting up to life again in the midst of the very rock garden itself, of all places impermissible and improbable. For the rich must have their money's worth in show; culture will not give it them, nor rarity, nor interest of the plants themselves: better a hundred yards of Arabis than half a dozen vernal Gentians. So now their vast rock-works are arranged like the pattern of a pavement: here is a large triangle filled neatly with a thousand plants of *Alyssum saxatile*, neatly spaced like bedded Stocks, and with the ground between them as smooth and tidy as a Guardsman's head; then, fitting into this, but separated by stone or rock, more irregular great triangles of the same order – one containing a thousand Aubrietia 'Lavender,' and the next a thousand *Lithospermum prostratum*. But nothing else; neither blending nor variety – nothing but a neat unalloyed exhibit like those on 'rock-works' at the Chelsea Show. But what a display is here! You could do no better with coloured gravels. Neat, unbroken blanks of first one colour and then another, until the effect indeed is sumptuous and worthy of the taste that has combined such a garden. But 'garden' why call it? There are no plants here; there is nothing but colour, laid on as callously in slabs as if from the paint-box of a child. This is a mosaic, this is a gambol in purple and gold; but it is not a rock garden, though tin chamois peer never so frequent from its cliffs upon the passer-by, bewildered with such a glare of expensive magnificence. This is, in fact, nothing but the carpet-bedding of our grandfathers, with the colour-masses laid on in pseudo-irregular blots and drifts, instead of in straight stretches ...[17]

The 'tin chamois' identify Friar Park[18] unmistakably, though the chamois with which Crisp ornamented the 30-foot scale model of the Matterhorn which was the climax of his rock garden were of cast iron. Much else in the diatribe is of dubious relevance: Crisp's collection of alpines numbered nearly 3,000 species and varieties, including, according to Robinson, more rarities than in any other collection, and his skills as an alpinist enabled him to go into partnership with the nursery of John Waterer and Sons.

But Farrer's main point was the massing of colour. Crisp was able to quote Henri Correvon (also an admirer of Ellen Willmott's rock garden) in defense of 'massed groups, broad belts which one can only describe by "as in Nature"',[19] but a quotation from Robinson would have been more to the point:

> If we have, for instance, fifty plants of the common Gentianella (Gentiana acaulis), it is better to make one or more large groups or carpets of them than to scatter them all over the rock garden one by one. Better still would this practice be with the Vernal Gentians, which, being a slow grower and smaller, is more likely to be exterminated by rapid-growing neighbours. Weeds and 'interlopers' are seen at a glance in such groups or carpets, and may be promptly dealt with ... Often, in consequence of not adopting the natural grouping or massing system, people are tempted to plant Ivies, Periwinkle, Clematis, and like plants, so as to hide the bareness of the ground; but bare ground may be easily covered by following the right system with true alpine flowers ... It was the wretched effect of the old mixed planting which brought forth the 'massing' system, with its original, simple, and telling beds. The great want in our gardens, not only in the rock garden, but in all the departments, is the massing and grouping system, minus geometry.[20]

The irony, which Farrer emphasised, was that Robinson had long been noted for his attack on the High Victorian bedding system, which had been characterised by the very massing of colours which he was advocating for the rock garden. Indeed, this recommendation is absent from the earlier editions of *Alpine Flowers for English Gardens* (1870 to 1879); but it appears in the first edition of the *English Flower Garden* (1883), and in the later editions of *Alpine Flowers* it is given emphasis by new illustrations, including one of Friar Park. This swing toward colour planning coincided with his increasing tolerance of colour schemes in the flower garden, under the influence of Gertrude Jekyll; and it is worth noting that early responses to Jekyll's writings on colour planning made the same sort of points that Farrer made − e.g. that it was a covert return to the principles of the bedding system.[21]

In repudiating the practice of the Robinson era, Farrer was explicitly rejecting not only the older tradition of geologically accurate rockwork, but also the more recent one of colour planning. What did he offer in their stead? Farrer was more precise in his condemnations than in his recommendations, in which he tended to lay stress on imagination and a true art that was not reducible to narrow rules. He is most often associated with scree or moraine planting as a more suitable medium than exposed rocks for the cultivation of alpines, but in this he was hardly being original. P. Rosenheim, in a remark made at the Rock Garden Conference in 1936, suggested that Farrer had been influenced by Sundermann of Lindau, who had published a pamphlet on scree-making as early as 1889;[22] but its English origins can be traced back at least to C.P. Peach, who wrote in the 1870s that alpine flora grew not on rocks themselves, but in 'the natural dells formed by the disintegration of rock'.[23] Farrer used the term 'moraine' for this sort of planting system, even though he acknowledged that what was being imitated was 'the high stone-slides of the Alps' instead of 'the pebbly, gravelly ridges of chaos' that the term properly indicated; this use of a term while publicly identifying it as a misnomer in itself indicates an already accepted tradition.[24] Farrer undoubtedly helped to popularise moraine or scree planting on a wide scale, but on this point his originality is more to be sought in his diminution of the need for rocks; he praised E.A. Bowles's rock garden for its virtual concealment of the rocks, and advised the beginning gardener that 'there need not be more than one' noble boulder in the garden. That rock garden was best which most resembled 'the unharvested flower-fields of the hills – effortless, serene, and apparently neglected'.[25]

Farrer's positive recommendations make greatest sense when seen against the background of the Edwardian debate between formal and informal design in the garden. Advocates of the formal condemned houses that appeared as if dropped from the sky onto a blank lawn; a house needed an architectural setting. Advocates of the informal tended to follow Robinson in invoking the concept of nature; but with this Farrer would have nothing to do. His characteristic rhetoric was more reminiscent of Loudon in his insistence on the necessity of art:

> For wildness without law is chaos; composed with regard to rhythm, it is the loveliest example of chained force that art can offer. But it is not 'natural': this is the word, the fatal word, that has deluded so many into thinking that rock-garden building is easy, because you just drop the stones all over the place, and it comes right ... 'Natural' in the garden is another word for laziness, ignorance, recklessness; nothing in the garden ... can possibly be natural. For the garden is a creation of art; and art is synthetized nature indeed, but governed by other (though no harder) rules than those that govern the constructions of the hills and forests. Drop all thought of the word 'natural' ...[26]

By the Edwardian period, the Japanese garden was becoming popular in England, though Farrer poured scorn on its characteristic English version; but it offered a model for informal design in the curtilage of the house, and Farrer, who repeatedly praised the artistic merit of true Japanese gardens (and their irreducibility to easily followed rules), looked forward to the adoption of a genuine Japanese influence:

> English gardening will take its next great leap forward when the R.H.S. prevails on Mr Lutyens to show a model garden, and on Mr Yoshio Markino to collaborate with him in compiling a living proof that a rock-garden can be made very beautifully to go in alliance with house and garden.[27]

It didn't happen, of course; Lutyens, and the other architects of his background, had too strong a commitment to formal design as a necessary accompaniment to the house. But for all that Farrer's ideals were difficult to reduce to specific recommendations, his rhetoric was an inspiration for the next generation of gardeners. It may be doubted with what degree of pleasure he would have responded to the great rock gardens of the interwar years – he would probably have been more enthusiastic about the spread of 'alpines without a rock garden' since the last war – but they would not have emerged as they did without his influence.

REFERENCES

1   Reginald J. Farrer, *The Rock-Garden* (T.C. & E.C. Jack, London, 1912), pp.1–2..
2   Reginald J. Farrer, 'Rock gardens and garden design', in *The Horticultural Record*, ed. R. Cory (J. & A. Churchill, London, 1914), p.3.
3   For the history of 19th-century rock gardens, see Brent Elliott, *Victorian Gardens* (Batsford, London, 1986), pp.46–8, 94–9, 176–8, 187–92; see also Graham Stuart Thomas, *The Rock Garden and its Plants* (Dent, London, 1989).
4   John Claudius Loudon, *The Suburban Gardener and Villa Companion* (London, 1838), pp.141–2.
5   C.P. Peach, 'Artificial rockwork', *Journal of Horticulture*, 30 (1876), p.152.
6   William Robinson, *Alpine Flowers for English Gardens* (London, 1870), esp. pp.9–14 for Backhouse's instructions.
7   On Richard Potter, see R.W. Wallace, 'The rise of the modern rock garden', in *Rock Gardens and Rock Plants: Report of the Conference held by the Royal Horticultural Society and the Alpine Garden Society*, ed. F.J. Chittenden (Royal Horticultural Society, London, 1936), pp.31–32.
8   Cory, *The Horticultural Record*, p.9.
9   Farrer, *The Rock-Garden*, p.1.
10   Cory, *The Horticultural Record*, p.10.
11   Farrer, *The Rock-Garden*, pp.8–9.
12   Cory, *The Horticultural Record*, pp.9–10.
13   Reginald J. Farrer, *The English Rock Garden*, (2 vols, TC & E.C. Jack, London, 1919), vol.I, pp.xvii-xviii.
14   See Mea Allan, *E.A. Bowles and his Garden at Myddleton House, 1865–1954* (Faber, London, 1973), pp.120–125; and Audrey le Liévre, *Miss Willmott of Warley Place* (Faber, London, 1980), pp.159–161.
15   Correvon in *The Garden*, 19 August 1905, pp.105–6; Ellen Willmott, *Warley Garden in Spring and Summer* (Bernard Quaritch, London, 1909).
16   *Gardening Illustrated*, Supplement to 23 May 1914. Hooper Pearson, in a letter to Bowles on 16 May 1914, as the storm was bursting, said that 'Robinson ... has distinguished himself lately by putting very extraordinary things into Gardening Illustr. and some of his readers have written me to say that they have no further use for his paper.' Letter in the Bowles/Crisp file, RHS Lindley Library.
17   Edward Augustus Bowles, *My Garden in Spring* (T.C. & E.C. Jack, London, 1914), pp.vii-viii.
18   While most of those who wrote to Bowles after Crisp's attack claimed that they had taken Farrer's remarks as generalisations, and had not noticed an application to Friar Park until after Crisp had spoken, there is a peculiar comment in Hooper Pearson's letter of 16 May 1914 to Bowles: 'Your memory is at fault in thinking you drew any attention to Farrer's attack on Friar Park. All you mentioned was the episode at Mont Cenis' – i.e., Farrer's depiction of Clutton-Brock's poor mountaineering in *The Dolomites* (1912). Clutton-Brock responded in a letter to Bowles: 'As for Farrer, I confess I think he deserves all he gets ... I cannot help feeling & speaking with bitterness about Farrer ...' (28 May). Robinson's remarks about Farrer in his letters are comparably incensed: 'Mr Farrer's uncontrolled talk' (26 May 1914); 'It has arisen wholly from Mr Farrer's too facile rush of words, often without clear aim' (27 May 1914); 'Farrer's careless way of writing' (5 June 1914).

19 *Gardening Illustrated,* Supplement to 23 May 1914.
20 William Robinson, *English Flower Garden* (John Murray, London, 1883), pp.1vii-1viii.
21 *Garden,* 22 (1882), pp.470–1.
22 P. Rosenheim, note in Chittenden, *Rock Gardens and Rock Plants,* pp.18–19, F. Sundermann 'Alpenwiese und Gerollfeld' (1889).
23 Peach, 'Artificial rockwork'.
24 Farrer, *The Rock-Garden,* p.28.
25 Bowles, *My Garden in Spring,* pp.ix-x; Farrer, *The Rock-Garden,* pp.7–11.
26 Cory, *The Horticultural Record,* p.13. For the debate, see Elliott, *Victorian Gardens,* pp.196–202.
27 Cory, *The Horticultural Record,* pp.11–12.

6

PLANT COLLECTING TODAY ...
IN THE STEPS OF FARRER

A.D. Schilling

There are many paths a man can follow in order to leave a significant and lasting record of his life's achievements. He may paint a masterpiece, write books or explore remote corners of the world and return with his discoveries which he will share with others. Reginald Farrer – artist, author and plantsman, achieved all these things; he was an enigma in his own lifetime and remains so today over seventy years after his death. The mountain paths he trod in his unquenchable search for new plants were often hard and lonely, but many plantsmen today yearn to follow in his footsteps and experience the same joy of discovery which he recorded in such passionate prose.

As long ago as the first decade of the 20th century, Veitch suggested to E.H. Wilson that almost everything worthwhile had been discovered and introduced into cultivation. Twenty years earlier A.E. Pratt wrote: 'So little of this great world of ours is new to the explorer or naturalist that it becomes more and more difficult year by year to find unworked fields'. I wonder what these men would think today at the close of the same century, when man has walked on the moon and explored even further and still new plants are being discovered and introduced to our gardens? Although we can argue that 'The golden age of plant-collecting' has passed, there is still much of scientific discovery waiting for the experienced man with sufficient enthusiasm to follow the mountain paths and to recognise the new or unusual.

However knowledgeable or sharp-eyed he may be, he will also require a certain amount of luck, for the ways are many and complex. The main irony of today's plant collecting rationale is that connected with time. Although jet travel has shortened journeys from weeks or months to a few hours, collectors no longer have the means or the inclination to stay in the field for extended periods. We live in a period of impatience and must therefore pay the price.

In 1919 when Farrer and Cox were planning their first expedition to Asia, Farrer had ideas to explore Nepal but this proved to be politically impossible. Today Nepal is an intensively explored country, a tourist attraction no less, and open to all. What purple passages would Farrer have composed if he had seen, as I and many others have, the sunset on Everest, the 3½ miles deep Kali Gandaki gorge, the golden flowered rosettes of *Primula aureata* in the Langtang Valley, or the blue flowered cushions of *Gentiana depressa* around the Sherpa villages of the Khumbu? We can only guess.

Following early boyhood days spent exploring the botanical wealth of the Yorkshire Dales he graduated to the European Alps often in the company of that great gardener E.A. Bowles.[1] Those of us who wish to follow in his wake may readily do so; the paths of the Dolomites may be steep, but the way is open and the plants which he loved and wrote about are there to be enjoyed amidst the mountain majesty.

Let us presume, however, that one wishes to feel the satisfaction that Farrer experienced amidst the mountain ranges of China and Burma. His trail can be retraced to some degree by exploring British gardens and 'plant hunting' some of his original introductions which still flourish in cultivation. One can 'play the game' in many ways, by observing the plants in general terms, or, perhaps more meaningfully, by concentrating on his treasures in their actual order of discovery.

9. *Geranium farreri*

In 1914, in the company of William Purdom he explored in north-western Kansu and, although the province is less botanically rewarding than Yunnan and Sichuan it bore him fruitful results. *Acer davidii* (F.351) can still be found in British gardens and the tree in Clapham village, on the site which was once Farrer's Craven Nursery, is almost certainly from this collection.

Gentiana Farreri.

A.M., R.H.S. 1919.　　F.C.C. 1920.　　Silver Medal, 1921.

Strong Plants - 3/6 and 5/- each.

W. WELLS, Jun.,

Hardy Plant Nurseries. Merstham, Surrey.

10. Advertisement for *Gentiana farreri* from *Nature* 1921

If one wishes to see the lovely late-summer flowering *Anemone vitifolia* var. *tomentosa* (F.436) one should make a pilgrimage to the famous chalk garden of Highdown[2] in West Sussex where many of Farrer's original plants from Kansu still thrive. It was here in 1985 that I 're-discovered' *Jasminum humile* var. *farreri* (F.867), for it was originally described from herbarium material and not thought to have been established in cultivation. Such are the excitements of *ex situ* plant hunting expeditions! Highdown is rich in Farrer introductions and here can be seen *Viburnum farreri* (F.13), *Buddleja alternifolia* (F.100), *Deutzia longifolia* var. *farreri* (F.109), *Potentilla fruticosa* var. *farreri* (F.188), *Carpinus turczaninowii* (F.331),

Euonymus phellomanus (F. 392) and much more besides. Alas, the notoriously difficult *Stellera chamaejasme* (F.93) has faded away.

If one yearns to see other of his Kansu plants such as *Geranium farreri, Rosa elegantula* forma *persetosa, Allium farreri, Clematis macropetala, Daphne tangutica* and *Meconopsis quintuplinervia* they can all be readily located in the specialist nursery trade. *Forsythia giraldiana* (F.388) grows at Kew, Edinburgh, Wakehurst Place and Windsor Great Park, whilst *Lonicera syringantha* and *Rodgersia aesculifolia* still flourish by the banks of Clapham lake on the Ingleborough Estate. The flannel-flowered *Buddleja farreri* (currently believed to be a large-leaved form of *B. crispa*) can be found against the wall in the Duke's Garden at Kew and *Viburnum betulifolium* (various Farrer collections) occurs at Edinburgh along with *Tilia laetevirens* (F.393), surely the rarest of all cultivated Limes.

Sadly, the interesting *Buddleja davidii* var. *nanhoensis* (F.424) from the river shingles of the Nan Ho seems to have been lost from cultivation, although incorrectly named plants frequently excite the eye either in gardens or within the pages of a nurseryman's catalogue. Even sadder is the loss of the true *Gentiana farreri* (F.217). All of the stocks now grown under this name are actually seed-raised hybrids which have gradually become 'mongrelised' over the years. There are no realistic short cuts to be made if purity is to be maintained, and until this lovely gentian is re-introduced this aggravating gardening gap must be lived with.

Compared with his Kansu collections, fewer of Farrer's Burmese plants have survived in cultivation as they are generally much less hardy. In spite of this, three which have adapted to the British climate have added considerably to the gardener's artistic palette. The scarlet ginger-lily, *Hedychium coccineum* (F.1098) is not a hardy plant in Britain, but it is well worthy of conservatory cultivation both for its flowers and foliage. Although this collection is probably not available from commerce it can be admired under glass at the Edinburgh Botanic Garden and those who are particularly struck by the beauty of this species may be interested to know that a Nepalese form of this ginger-lily which I collected east of Kathmandu approximately 15 years ago has proved to be extremely hardy and boasts larger orange-red flowers; it bears the clonal name 'Tara' and holds a First Class Certificate accolade.

The elegantly pendulous-branched 'coffin juniper' *Juniperus recurva* var. *coxii* (F.1407) is always admired by those who come across it, for it is a weeping conifer of great beauty; Cox described it as 'one of our best collections'. Good examples of this tree abound, but the finest I know grows at Exbury Gardens near Southampton in Hampshire. This great garden also boasts one of the few mature trees of *Picea farreri* (F.1435) in cultivation. Farrer collected it in the Feng-Shui-Ling valley of N. Burma and described it as 'a stately grey-green noble Spruce'. The name of this elegant species has only been described very recently and Keith Rushforth (who is one of the authors of the paper)[3] must be thanked for his efforts and success in propagating this extremely rare tree thereby establishing it more securely in cultivation.

11. *Hedychium coccineum* (Farrer 1098)

If one craves to see some of his Burmese rhododendrons in gardens one can do no better than visit the great woodland gardens of Britain, especially Windsor Great Park; there are also many Farrer numbers on rhododendrons at Edinburgh Botanic Garden. In these collections may be 're-discovered' such large-leaved gems as *Rhododendron arizelum* (F.863) and *R. basilicum* (F.873), the scarlet flowered *R. sperabile* (F.888) and *R. mallotum* (F.815), the yellows of *R. trichocladum* (F.876) and *R. caloxanthum* (F.937), the pinks of *R. glischrum* (F.887) and the mauves of *R. heliolepis* (F.878).

At Wakehurst Place in West Sussex can be found the dwarf hummocks of *R. campylogynum* (F.1046) and the large-leaved fragrant white-flowered *R. decorum* (F. 979) which Farrer termed 'a comely tree of wood edges'. This last named species, plus *R. fulvum* (F.874), can also be found by the gorge caused by the dramatic Craven Fault above the Clapham lake. Here at the foot of Ingleborough Farrer gardened a mini-Burmese woodland where bamboos grow poised on the bluffs above the turbulent waters of Clapham Beck. Although some of the plants thereabouts are definitely not Farrer's originals, many certainly are; others may well be his collections, but we can only speculate. Here, on a quiet day one can pause and almost expect to meet the man himself, such is the atmosphere of the spot.

Two other Burmese collections of his are worthy of special mention. Firstly, *Berberis coxii* (F.1030) which was recently 're-discovered' alive and well in the Sussex garden of Borde Hill and has since shown up elsewhere, thereby ensuring it a safer niche in horticulture. Secondly, one must mention the large spreading *Cotoneaster franchetii* var. *sternianus* which is frequently seen in arboreta throughout our islands; it even occurs by a dry stone wall close to a house in Newbycote which is so close to Clapham that one can assume it to be ex Craven Nurseries.

Finally, our pilgrimage can take us to the village gardens of his native Clapham for it boasts several plants of *Viburnum farreri* and *Potentilla fruticosa* var. *farreri*; the fine tree of *Pinus thunbergii* on the Craven Nursery site is both old enough and lovely enough to be linked to Farrer's Japanese journey, but to date I have traced no qualifying records to back the suggestion. In the grounds of Ingleborough Hall stands a fine specimen of the white flowered *Viburnum farreri* var. *candidissimum*, as well as a tree of *Juniperus recurva* var. *coxii*. If this isn't enough one can take oneself off up Ingleborough hill in quest of the purple saxifrage or the arctic sandwort which Farrer admired in his youth.[4]

From these selected notes it will be clear that we can readily tread in the footsteps of Farrer, either by walking his beloved Ingleborough, by strolling through gardens, by viewing his plantings in the gorge below Clapham Cave or by walking the same trails he followed in the Dolomites. Those of us who are more adventurous can probably visit Kansu province north east of the great lake of Koko Nor and the mighty peak of Amne Machin, but current politics prohibit any chance to realise the remote regions of Upper Burma where Farrer struggled against the savage climate in his passionate quest for new plants; a quest which ultimately cost him his life.

Little or no seed came home from his last journey – just herbarium specimens. Even his frying pan and boots were collected up and despatched by his loyal Gurkha orderly, but not the real fruits of his last labours. In spite of this we have numerous other fine plants to remind us of his earlier endeavours. Every time we come across a flowering specimen of *Viburnum farreri* on a sunny winter's day, or pause to admire the pendant racemes of *Buddleja alternifolia* in June, we acknowledge his Asian achievements. When we chance upon a plant of *Saxifraga hirseriana* 'Gloria' or *Wahlenbergia serpyllifolia* 'Major' in a nurseryman's frame we find ourselves appreciating the plants he selected closer to home. We are reminded of the great 'green legacy' he transmitted to posterity; it is easy to walk in his footsteps – after all he has left us such clear footprints.

Even what has been lost to us gives pause for romantic thought. Somewhere high on a far-off Burmese mountainside *Primula agleniana* flowers unseen by European eye and awaits re-discovery and introduction to our gardens.

In the early summer of 1987 I joined a Kew/Edinburgh/Kunming botanical expedition[5] to north west Yunnan, a region which lies close to the eastern borders of Chinese Tibet and Northeast Burma. After several weeks in the Yulong Shan mountains (better known as Lichiang) we moved westwards to botanise and explore the vast forest ridges which rise above the deep river gorges of the Mekong and Yangtse. This region, often referred to as part of 'the Tibetan marches', is a massive dissected plateau savagely riven by these two great rivers as well as others further to the west including the Salween and N'mai-hka. Our expedition was all too short, but we saw many plants to satisfy our eyes including the violet velvet flowers of the exquisite *Omphalogramma vincaeflorum*, a plant which caused Farrer to become 'half paralysed with excitement' when he first saw it. Such passages of purple prose prompted Cox to describe them as 'Farrer's trick of over-emphasis'.

As we stood there close to the Burmese border it occurred to us that we were no more than 100 miles from Farrer's grave at Nyitadi; not quite in his final footprints but closer than many men have been in recent years. Our country at Li-ti-ping was very similar to his, and the excitement we experienced was the same excitement which motivated him in Asia's mountain wilderness. In his book *The Dolomites*[6] Farrer wrote 'It depresses me when my companions have no idea as to where they are going or what they are seeing.' I and my companions, English, Scottish and Chinese alike, all knew exactly where we were and precisely what we were seeing that June day on the roof of the world close to Farrer's final resting place.

In conclusion it is appropriate to quote a line or two from Euan Cox's *Plant Hunting in China*[7] which although published 45 years ago still rings true today: 'The finding of new plants although exciting and satisfying is only one angle of a plant collector's job. To extend the range of species already known from other areas or to provide good material throwing light on variations is as important in the science of phytogeography as the discovery of a new species... Let us hope that the future will not repress the desire of men to breast again the high hills in search of plants.' Happily the quest for plants continues today, and there are still stories waiting to be told.

REFERENCES

1   See also the article by le Liévre.
2   See also the article by McLean.
3   Keith Rushforth, in Notes Royal Botanic Garden Edinb., 38(1) (1980), pp. 129–36.
4   See also the article by Roberts.
5   Christopher Grey-Wilson and A.D. Schilling, 'Botanising in mountain Asia', in *Plant Hunting for Kew*, ed. F. Nigel Hepper (H.M.S.O., London, 1989), pp.85–102.
6   Reginald J. Farrer, *The Dolomites* (Adam and Charles Black, London, 1913).
7   Euan H.M. Cox, *Plant Hunting in China* (Collins, London, 1945).

7

PLANT NAMES COMMEMORATING REGINALD FARRER:
A Bibliographical Record

William T. Stearn

Reginald John Farrer (1880–1920) enriched European and North American gardens with notable hardy Chinese plants and discovered many plants new to science, of which 29 bear the commemorative epithet *farreri* or *farreriana*; other species with the epithets *purdomii* and *coxii* honour his travel companions. This achievement is remarkable in that, by the amount of his collecting, the number of plants introduced and herbarium specimens collected, as also the length of time spent in the field, Farrer never rivalled the three great 20th-century British collectors of plants in Eastern Asia, George Forrest (1873–1932), Ernest Wilson (1876–1930) and Frank Kingdon-Ward (1885–1958). Nevertheless his accounts of his travels in search of plants stand as lasting contributions to the literature of plant-collecting. His horticultural contribution would have been much greater but for the 1914–1918 World War. Farrer's seeds came to Britain from China when too many British gardeners were serving in the armed forces, or were so totally engaged in growing food, for this precious hard-won material to receive the skilled attention it merited. The hardiness of those successfully raised, for which E.H.M. Cox's *The Plant Introductions of Reginald Farrer* (1930) provides a detailed account, derived from their provenance. Wisely Farrer chose Kansu (Gansu), the bleak north-western province of China, as likely to yield plants of sure hardiness in Britain. Wisely too he chose a steady, even-tempered, experienced Kew-trained north-country gardener, William Purdom (1880–1921), as companion and assistant. Kansu was not botanically virgin territory; indeed publications by Carl Johann Maximowicz based on herbarium collections by Russian collectors in that region, notably his *Flora Tangutica* (1889), probably directed Farrer's attention to its horticultural potentialities, for he surveyed an immense range of botanical works when compiling his encyclopaedic *The English Rock-Garden* (2 vols, 1919). The Russian zoological and botanical explorer Nicolai M. Przewalski made collections in 1872, 1873, 1880 and 1884 in the very area chosen by Farrer, gathering about 15,000 botanical specimens representative of some 1,700 species and introducing a few into cultivation at St Petersburg (Leningrad). Later, in 1884, another Russian collector, Grigori N. Potanin, and his wife made very large collections in the area, with their headquarters at Siku in Kansu. Farrer and Purdom likewise made their headquarters in Siku. It is worthy of note that both Przewalski and Potanin had been Russian army officers undeterred by potential lawlessness which Farrer and Purdom feared; on one expedition Przewalski was accompanied by two army officers, three ordinary soldiers, five cossacks, a taxidermist and interpreters!

Farrer and Purdom thus had the prospect before them of introducing into British gardens many plants already made known by these industrious Russian collectors but not of discovering many new to science. Nevertheless investigation at the Royal Botanic Garden, Edinburgh, of living and herbarium material from their 1914–1915 Kansu expedition, as also from Farrer's fatal 1919–1920 expedition to Upper Burma, revealed a surprising number of plants needing descriptions and names. This, however, is not really astonishing for such a rugged area; thus intensive work in recent years on even so accessible a mountainous area as Greece has brought to notice many new species. Botanists naming new species can provide them with a great diversity of specific epithets, some referring to features of the plant itself or its place of collection but others gratefully commemorating the collector. As regards such honorific names Farrer has been fortunate, together with his travel companions William Purdom and E.H.M. Cox. Some names commemorating Farrer have passed into synonymy but most of them stand as accepted specific or varietal designations. The following list supplies much information not available when Euan H.M. Cox (1893–1977) published his *The Plant Introductions of Reginald Farrer* (1930), a work limited to 500 copies. The type-localities below are quoted direct from the original descriptions which mostly incorporate Farrer's field-notes. As regards the transcription of Chinese place-names into the Roman alphabet, Farrer considered that 'The Wade system is quite the finest bad-form joke that has yet been perpetrated on humanity' and he used his own phonetic renderings and translations, e.g. 'Wolvesden' for Long Shih Tang. T.F. Wade and H.A. Giles, both professors of Chinese, based their system of romanizing Chinese on the northern dialect. Chinese pronunciation varies greatly from place to place and the Wade-Giles system has now been superseded by the Chinese Government's modern system of romanization known as Pinyin. Thus Kansu is now rendered Gansu.

LIST OF NAMES COMMEMORATING FARRER

In the following list the heading in bold type to each entry gives the botanical name currently accepted taxonomically and nomenclaturally followed by the name of the botanist (or botanists), responsible for its publication and the place, a periodical or book, where it was published. The abbreviation 'Syn.' (Synonym) is followed by another name for the same plant not according with current classification. The paragraph 'Type' contains the information provided in the first description as to the material on which that description was based, this often quotes Farrer's descriptive note accompanying his herbarium specimen used.

Allium cyathophorum Bureau & Franchet var. **farreri** (Stearn) Stearn in Curtis's Bot. Mag. 170: N.S. t.252 (1955).
Syn. *A. farreri* Stearn in J. Bot. (London) 67: 342 (1930). Liliaceae (Alliaceae).
Type: 'China (Siku) "165 Allium (narrow leaf) Siku 5500, June 29, 1914" Farrer in Herb. Roy. Bot. Gard. Edinburgh sed e speciminibus cultis descriptum' (Stearn, loc. cit. 1930).

Amitostigma farreri Schlechter in Fedde, Repert. Sp. Nov. 20: 378 (1924).
Syn. *Orchis farreri* (Schlechter) Sóo in Ann. Mus. Nat. Hungar. 26: 348 (1929). Orchidaceae.
Type: 'West-Tibet: Abundant in a little marshy stretch of lawn at the valley-head, down on the Chinese side of the Chawchi-Pass, 11,800–12,500 ft. – R. Farrer No. 1856, Aug. 1920' (Schlechter, loc. cit.).

Aster farreri W.W. Smith & J.F. Jeffrey in Notes R. Bot. Garden, Edinburgh 9: 78 (1916); Cox, Introd. 30 (1930).
Syn. *Erigeron farreri* (W.W. Smith & J.F. Jeffrey) Botschantzev in Notulae Syst. Inst. Bot. URSS 21: 341 (1961). Compositae (Asteraceae).
Type: '"Very handsome and sporadic in the higher fields and alps of Tibet, in hay grass along with No. 173 but not ascending so high. 12th August 1914; East Tibet, near Kansu frontier" Farrer & Purdom No. 174' (Smith & Jeffrey, loc. cit.).

Berberis farreri Ahrendt in J. Linn. Soc. London, Bot. 57: 192 (1961). Berberidaceae.
Type: 'West Kansu: Lo sin, Lan chan, 9 July 4138 (K). seed only, Farrer 318.
Cultivated: from Farrer 318, fl. 5 June 1939 (*Type*, O), ex H.G. Hawker, Stode, Devon, comm. A.B. Jackson' (Ahrendt, loc. cit.).

Buddleja crispa Bentham var. **farreri** (I.B. Balfour & W.W. Smith) Handel-Mazzetti, Symb. Sin. 7: 947 (1936).
Syn. *B. farreri* I.B. Balfour & W.W. Smith in Notes R. Bot. Garden, Edinburgh 9: 84 (1916); Curtis's Bot. Mag. 150: t.9027 (1924); Bean, Trees & Shrubs, 8th ed., 1: 453 (1970). – *B. tibetica* W.W. Smith var. *farreri* (I.B. Balfour & W.W. Smith) Marquand in Kew Bull. 1930; 205 (1930). Loganiaceae.
Type: '"This noble bush of 4–6 ft., with ample boughs of huge flannelly foliage, hugs only the very hottest and driest crevices, cliffs, walls, and banks down the most arid and torrid aspects of the Ha Shiu fang (about Siku), and the baking

stony defiles of the Feng S'an Ling (S. side). It does not range northward, and the flowering specimen was gathered from a strange outlying colony at the edge of subalpine coppice below Chago, in the Satanee Valley on 8th May. These magnificent thyrses appear before the leaves, which afterwards unfold to hide all trace of them: they suggest a glorified *Veronica Hulkeana* on a big scale, and have the most delicious scent of raspberry ice. Kansu, West China." Farrer and Purdom. No. 44 in Herb. Edin.' (Balfour & Smith, loc. cit.).

Bulbophyllum farreri (W.W. Smith) Seitenfaden in Dansk Bot. Ark. 29: 212 (1974).
Syn. *Cirrhopetalum farreri* W.W. Smith in Notes R. Bot. Garden, Edinburgh 13: 196 (1921). Orchidaceae.
Type: 'Upper Burma. Cultivated in Royal Botanic Garden, Edinburgh, from plants collected by the late Mr Reginald Farrer' (Smith, loc. cit.).

Callianthemum farreri W.W. Smith in Notes R. Bot. Garden, Edinburgh 9: 90 (1916); Cox, Pl. Introd. 32 (1930). Ranunculaceae.
Type: '"On cool peaty ledges of the Satanee range, 8,000–10,000 ft. Flowers 6th–15th May xx Kansu, West China" Farrer and Purdom. No. 70' (Smith, loc. cit.).

Cirrhopetalum farreri W.W. Smith (1921). See *Bulbophyllum farreri* (W.W. Smith) Seitenfaden.

Codonopsis farreri Anthony in Notes R. Bot. Garden, Edinburgh 15: 181 (1926). Campanulaceae.
Type: 'Burma. "Chimili Valley. Plant attaining 3–4 ft. Twines elegantly up the culms of the bamboos in the uppermost woodland. Flower yellow, heavily veined and flushed in the lobes with claret colour. Alt. 11,000 ft. 31st July 1919" Farrer 1144' (Anthony, loc. cit.).

Cotoneaster farreri Klotz in Wiss. Zeitschr. Univ. Jena, Nat. Wiss. 21: 1006 (1972). Rosaceae.
Type: 'Burma boreo-occidentalis, Mokuji-Pass, 3100 m. -22-Flower unknown; an erect bush of 2–3 feet, in the openest places of the alpine coppice, not common here-22-; leg. R. Farrer 8. VIII. 1926 [sic...] no. 1830(E)' (Klotz, loc. cit.).

Cremanthodium farreri W.W. Smith in Notes R. Bot. Garden, Edinburgh 12: 202 (1920); R. Good in J. Linn. Soc. London, Bot. 48 276 (1929); Cox, Pl. Introd. 62 (1930). Compositae (Asteraceae).
Type: '"Upper Burma: Chimili Alps. Alt. 12,500–13,000 ft. Local but locally

abundant in dips and dells of the longer high-alpine grass. A stately species with pendulous globular-looking flowers of pure white, that deepen to dark claret-colour as they die. Aug. 1919" Farrer. No. 1178' (Smith, loc. cit.).

Cryptochilus farreri Schlechter in Fedde, Repert. Sp. Nov. 20: 384 (1924). Orchidaceae.

Type: 'West-Tibet: Below the brags of Shin-Hong, 9,500 ft. R. Farrer No. 1649, June 1920' (Schlechter, loc. cit.).

12. Farrer's painting of *Cypripedium farreri*:
"Cypripedium Sweetlips Siku. July 1/14"

Cyananthus lobatus Bentham var. **farreri** Marquand in Kew Bull. 1924: 247 (1924); Cox, Pl. Introd. 61 (1930). Campanulaceae.

Type: 'Western China (exact locality unknown) Farrer' (Marquand, loc. cit.). According to Cox, this was collected and introduced into cultivation as Farrer 1220 from Upper Burma, growing there at 13,000 feet.

Cypripedium farreri W.W. Smith in Notes R. Bot. Garden, Edinburgh 9: 102 (1916).

Syn. Cypripedium species Farrer in Gard. Chron. III: 57: 258 (1915). Orchidaceae.

Type: '"Rare in the deep limestone gorges above Siku, under the shadier wall of the cliffs. June. Petals etc. greeny-yellow, lined maroon; lip of waxy cream, lined internally, pulled into a series of vandykes at the mouth and very glossy and fragrant. Kansu, West China". Farrer and Purdom. No. 155' (Smith, loc. cit.).

Deutzia longifolia Franchet var. **farreri** Airy Shaw in Curtis's Bot. Mag. 161: t.9532 (1938).

Syn. "*D. albida* Batal." sensu Cox, Pl. Introd. 35 (1930) non. Batalin.

Type: 'W. China; eastern Kansu; "abounds about Mö-ping, where all the coppiced slopes are a surf of snow". *Farrer* 109 in Herb. Kew' (Airy Shaw, loc. cit.).

13. Part of the herbarium specimen of *Gentiana farreri* from the Royal Botanic Garden, Edinburgh. "F. 807A Gentiana Farreri, Balf. Very abundant and of outstanding beauty in the high finer turf of the Da Tung alps, from 10, to 14,00 ft extending to the grassy passes between Tien Tang & Ping fan, where just on the crests, it simply sheets the earth with azure on Sept 15th & 16th 1915"

Gentiana farreri I.B. Balfour in Trans. Bot. Soc. Edinburgh 27: 248 (1918); Cox, Pl. Introd. 36 (1930); Marquand & Ballard in Curtis's Bot. Mag. 147: 8874 (1938).

Syn. *Gentianodes farreri* (I.B. Balfour) A. Löve & D. Löve in Bot. Notiser 125: 257 (1972). Gentianaceae.

Type: 'Kansu. Jo-Ni alps. Farrer & Purdom 1914 specimens of this species were not brought back by Farrer and my description is based upon living plants which flowered in the Royal Botanic Garden [Edinburgh] in August 1916' (Balfour, loc. cit.). In fact the seeds were gathered by Chinese collectors for Farrer.

Geranium farreri Stapf in Curtis's Bot. Mag. 151: t.9092 (1926); Cox, Pl. Introd. 37 (1930); Yeo, Hardy Geraniums 101 (1985). Geraniaceae.

Type: 'China: Min-shan on the border of Kansu and Szechuan, 3,900–4,500 m' (Stapf, loc. cit.).

Iris farreri Dykes in Gard. Chron. III. 57: 175 (1915). Iridaceae.

Type: 'No. F.325 "Abundant by the upland tracks and in open places of the Min S'an, not below 9,000 ft. nor above 10,000 ft. July 20 (lingering)"' (Dykes, loc. cit.).

Isometrum farreri Craib in Notes R. Bot. Gard., Edinburgh 11: 250 (1970) Gesneriaceae.

Type: 'S. Kansu. Very general at low elevations on rather cool rocks and very steep banks of cool clammy soil that grows a fine film of moss. Flowers a pretty shrimpy pink with a bronzy tone. Farrer et Purdom, 262. Fl. Aug. 28' (Craib, loc. cit.).

Jasminum humile L. forma **farreri** (Gilmour) P.S. Green in Bean, Trees & Shrubs, 8th ed., 2: 465 (1973).

Syn. "*J. giraldii*" sensu Cox, Pl. Introd. 66 (1930) non Diels. – *J. farreri* Gilmour in Curtis's Bot. Mag. 47: t.9351 (1934). Oleaceae.

Type: 'This alternate-leaved Jasmine was collected by Farrer on April 25th, 1919, at 2286 m on an "open arid outcrop of limestone ridge just below the Dâk Bungalow", Hpimaw Hill, Upper Burma (F.867)' (Gilmour, loc cit.)

Lilium duchartrei Franchet in Nouv. Arch. Mus. Hist. Nat. Paris II. 10: 90 (1887); Woodcock & Stearn, Lilies of World 212 (1950); Haw, Lilies of China 113 (1986).

Syn. *L. farreri* Turrill in Gard. Chron. III. 66: 76 (1919), Curtis's Bot. Mag. 146: t.8847 (1920). – *L. duchartrei* var. *farreri* (Turrill) Grove in Gard. Chron. III. 78: 69 (1925); Cox, Pl. Introd. 40 (1930). Liliaceae.

Type: 'grown from seed collected by Mr Farrer' (Turrill, 1919).

Lonicera farreri W.W. Smith in Notes R. Bot. Garden, Edinburgh 9: 110 (1916). Caprifoliaceae.

Type: '"A little frail bush of 3 feet or so, with *flattened* outspread sprays from which the rosy bugles hang − a plant of unique charm. There are larger and coarser approximations to this, in the lower Alpine coppice of the Satanee Range, from 7,000−8,000 ft.; but of this form I have only seen two certain plants − one just above Chago by the pathside, and the other on a cliff above a torrent in a deep ghyll behind Ga-hoba. 6th May, 10th May 1914. Kansu, West China." Farrer and Purdom. No. 46.' (Smith, loc. cit.).

Nomocharis farreri (W.E. Evans) Harrow in New Fl. & Silva 1: 76 (1928); Sealy in Curtis's Bot. Mag. 162: t.9557 (1939), Bot. J. Linn. Soc. London 87: 300 (1983); Haw, Lilies of China 148 (1986).

Syn. *N. pardanthina* var. *farreri* W.E. Evans in Notes R. Bot. Garden, Edinburgh 15: 20 (1925); Cox, Pl. Introd. 70, plate facing p.44 (1930). Liliaceae.

Type: 'Upper Burma. "Hpimaw Pass, alt. 10−11,000 ft. In a wild state the plant is even finer [than when cultivated] and abounds in millions over the open alp-slopes on either side of the Pass and below it, descending even into light glades of bamboo, and little dells on the woodland edges. Shape, colour and spotting variable; flowers from 1−10 a plant, and a spectacle of unsurpassed beauty and charm. Fullest bloom will be about June 25th" June 1919. R. Farrer. 1031! Cultivated specimens from the same neighbourhood were sent by Mr Farrer under the number 988! He states that the bulbs are eaten.' (Evans, loc. cit.).

Omphalogramma farreri I.B. Balfour in Notes R. Bot. Garden, Edinburgh 13: 23 (1920); W.W. Smith & Forrest in Notes R. Bot. Garden, Edinburgh 15: 255 (1927); Cox, Pl. Introd. 71 (1930). Campanulaceae.

Type: 'N.E. Upper Burma. Hpimaw and Chimili High Alps. Farrer and Cox 1053, 1169' (Balfour, loc. cit.).

'We append the original notes of Farrer on the two numbers quoted in the original description:-

"N.E. Burma: − Hpawshi Bum. Alt. 12,300 ft. In the light dwarf high alpine scrub. Foliage even more violoid than in *O. Viola-grandis*. The original diagnosis [i.e., of *Delavayi* with which Farrer identified it] does not do justice to this feature, though I cannot otherwise separate the plant from *O. Delavayi* as I should like. June 24th 1919." R. Farrer. No. 1053.

"Hpimaw and Chimili Alps. Alt. 12,600−12,800 ft. Among high-alpine brushwood and even in higher alpine hayfields and forming wonderful masses in alpine marshes and rill-sides. Flowers dark purple. Leaves violoid. (Cf. No. 1187, a very ugly and quite distinct twin.) June 24th 1919." R. Farrer. No. 1169' (Smith & Forrest, loc. cit.).

In the *Fl. Reipubl. Pop. Sinicae* 58 ii (1990) Chi-Ming Ho considers *O. farreri* synonymous with *O. delavayi* (Franchet) Franchet (1898).

Onosma farreri I.M. Johnston in J. Arnold Arb. 32: 345 (1951)

Syn. *O. sinicum* Diels var. *farreri* (I.M. Johnston) W.T. Wang & Y.L. Liu in Acta Phytotax. Sinica 18: 70 (1980). Boraginaceae.

Type: "China (southern Kansu) betw. Kiai Chow and Wen Hsien, very abundant on dry bank, Apr. 26. 1914, *R. Farrer* 3 (Type, Gray Herb.) '. . . "on all the torrid banks in the torrid region of the Blackwater and the Nan Ho" ' (Johnston, loc. cit. 346).

Paraquilegia anemonoides (Willdenow) Ulbrich in Fedde, Repert Sp. Nov. Beih. 12: 369 (1922)

Syn. *Aquilegia anemonoides* Willdenow in Mag. Ges. Naturf. Freunde (Berlin) 5: 401 (1811) – *Isopyrum grandiflorum* Fischer ex DeCandolle, Prodr. 1: 48 (1824) – *Paraquilegia grandiflora* (Fischer) J.R. Drummond & Hutchinson in Kew Bull. 1920: 156 (1920) – *Isopyrum farreri* auct. Edinb.; Farrer, On Eaves of World 1: fig. facing p.300, 2: 521 (1917), nomen nudum: Alpine Garden Soc. Bull. 1: 191, t.92 (1932). Ranunculaceae.

Farrer's plant belongs to a variety of this beautiful and variable species having lavender-blue erect flowers.

Orchis farreri (Schlechter) Sóo (1929). See *Amitostigma farreri* Schlechter.

Parnassia farreri W.E. Evans in Notes R. Bot. Garden, Edinburgh 13: 174 (1921). Saxifragaceae (Parnassiaceae).

Type: ' "Upper Burma:- Chimili Valley. Alt. 11,000 ft. Flowers white. Characteristic situations in the middle alpine zone 15.08.19" R. Farrer, No. 1211. Type !' (Evans, loc. cit.).

Picea farreri Page in Notes R. Bot. Garden, Edinburgh 38: 130 (1980).

Syn. *Picea* species, Cox, Pl. Introd. 71 (1930). Pinaceae.

Type: 'Upper Burma: Feng-Shui-Ling valley, 1919, Farrer 1435 (holo. E, cones and seeds only) . . . This spruce was first collected in the Feng-Shui-Ling Valley ('Valley of the Winds and Waters') of Upper Burma (approx. lat. 25°50'N, long. 98°30'E) by Reginald Farrer in 1919, under the number 1435 . . . A single tree from seed of *Farrer* 1435, planted in 1922, was located at Exbury Gardens, Hampshire, in 1976 (Page, loc. cit.).

Potentilla parvifolia Fischer ex Lehmann, Nov. Minus Cognit. Stirp. Pugillus 3: 6 (1831); Bean, Trees & Shrubs, 8th ed. 3: 337 (1976).

Syn. *P. fruticosa var. farreri* hort; Cox, Pl. Introd. 44 (1930). Rosaceae.

Type: 'Kansu. "This form being all of a pure golden yellow and making a glorious effect in great bushy jungles in open glades and along the fringes of woods". F.188' (Cox, loc. cit.).

Primula farreriana I.B. Balfour in Notes R. Bot. Garden, Edinburgh 9: 167 (1916); W.W. Smith & J.R. Fletcher in Trans. R. Soc. Edinburgh 60: 581 (1942). Primulaceae.

Type: '"Kansu. Ta-Tung Alps. Dark cold and damp gullies or tight cliff crevices in shade on calcareous or non-calcareous rock from 12,000–15,000 ft., very sweet." Farrer and Purdom. F.560. Primula No. 29. June-July 1915.' (Balfour, loc. cit. 168)

Rhodiola dumulosa (Franchet) Fu forma ***farreri*** (W.W. Smith) Jacobsen in National Cactus & Succ. J. 28: 5 (1973).

Syn. *Sedum farreri* W.W. Smith in Notes R. Bot. Garden, Edinburgh 9: 125 (1916). Crassulaceae.

Type: '"In the limestone screes at great elevations only, from the Min S'an Alps down to Thundercrown, at 12,000–14,000 ft. Kansu, West China. 28 Aug. 1914." Farrer and Purdom No. 238' (Smith, loc. cit.).

Rhododendron oreodoxa Franchet in Bull. Soc. Bot. France 33: 230 (1886); Rhododendron Society, Species of Rhododendron 285 (1930).

Syn. *Rhododendron reginaldii* I.B. Balfour in Notes R. Bot. Garden, Edinburgh 11: 114 (1919).

Type: 'S. Kansu. "Only seen above 9,000 ft. in one series of wooded or coppiced mountain glens on the 10,000 ft. range intervening between the main chains of Siku and Satanee, *not* (for once) on limestone, but on a red shale. A comely pyramidal bush or round-headed tree of 12–15 ft., exceedingly profuse, with lovely pale-pink flowers." Farrer, No. 63. May 12, 1914' (Balfour, loc. cit.).

Balfour used Farrer's Christian name for an epithet owing to the existence of the name *Rhododendron farrerae* Sweet, Brit. Flower Garden II.1:t.95 (1931) honouring Mrs Farrer of Blackheath, whose husband William Farrer, captain of the Hon. East Company's ship 'Orwell' brought the plant from China about 1828.

Rosa elegantula Rolfe in Kew Bull. 1916: 100 (1916); Bean, Trees & Shrubs, 8th ed. 4: 89 (1980).

Syn. *R. farreri* Stapf ex Cox, Pl. Introd. 49 (1930); Stearn in J. Bot. (London) 70, Suppl.: 19 (1932); Stapf in Curtis's Bot. Mag. 147: t.8877 (1938). Rosaceae.

Type: 'This striking rose came up in E.A. Bowles's garden among seedlings raised from hips collected by Reginald Farrer in southern Kansu, China' (Stearn, 1932, loc. cit.). This note refers to cv. 'Persetosa', called by Bowles 'Farrer's Threepenny-bit Rose'; owing to inflation the disappearance of the small threepenny coin has deprived that name of its aptness.

Saxifraga x farreri Druce in Bot. Exchange Club & Soc. Brit. Isles 2: 256 (1908), nomen nudum; Farrer, Engl. Rock-Garden 2: 270 (1919).

Type: 'Discovered by Mr Farrer on Ingleborough and showing evidence of both parents S. hypnoides and S. tridactylites' (Druce, loc. cit.).; 'found by me some years ago on the western face of Ingleborough' (Farrer, loc. cit.).

Sedum farreri W.W. Smith (1916). See *Rhodiola dumulosa* forma *farreri* (W.W. Smith) Jacobsen.

Trollius farreri Stapf in Curtis's Bot. Mag. 152: sub t.9143 (1928); Cox, Pl. Introd. 51 (1930); Doroszewska in Monogr. Bot. 41: 29 (1974); W.–T. Wang in Fl. Reipubl.Popul. Sinicae 27: 78, t.15 f.6 (1979).

Syn. *T. Kansuensis* (Bruhl) Mukerjee in Bull. Bot. Survey India 2: 106 (1960), Ranunculaceae.

'Published as T. pumilus in the Gardeners' Chronicle vol. LIX fig. 13 (1916) and in Farrer's English Rock-Garden II. t.45 (1919)' (Stapf, loc. cit.), hence collected in alpine grassy lawns near 'Wolvesden', Kansu as Farrer 137.

Viburnum farreri Stearn in Taxon 15: 22 (1966); Bean, Trees & Shrubs, 8th ed. 4: 697 (1980).

Syn. *V. fragrans* Bunge, Enum. Pl. China Bor. 33 No. 194 (1833); Cox. Pl. Introd. 52 (1930); Stapf in Curtis's Bot. Mag. 147: t.8887 (1938); non *V. fragrans* Loiseleur in Mordant de Launay, Herb. Amat. 7: t.466 (1824).

Type: 'Colitur in hortis ob odorem gratissimum' (Bunge, loc. cit.), i.e. cultivated in gardens around Peking (Beijing) for its very pleasing scent. It is unfortunate that a name long so well-known as *Viburnum fragrans* for this fine shrub has had to be rejected as a later homonym, but at least the substitute name *V. farreri* preserves an association with Farrer who, together with Purdom, introduced it into British gardens from the wild. The name *V. fragrans* Loiseleur (1824), which invalidates *V. fragrans* Bunge (1833), is synonymous with *V. odoratissimum* Ker-Gawler (1820), as are likewise the illegitimate names *V. dubium* Steudel (1841) and *Thyrsosma chinensis* Rafinesque (1838); all four names refer ultimately to a plant introduced from China into cultivation in England.

Wikstroemia pretiosa Domke in Notizblatt Bot. Gart. u. Mus. Berlin-Dahlem 11(105): 363 (1932) nomen nudum.

Syn. *Farreria pretiosa* [I.B. Balfour & W.W. Smith ex] Farrer, Eaves of World 2: 320 (1917), nomen nudum. Thymelaeaceae.

In *On the Eaves of the World* 1: 104 (1917) Farrer described how on 18 April 1914 at Ping Lu Tien, southern Kansu there 'shone like living sunlight patches of a radiant little golden yellow ground Daphne with flowers so much larger than are usual in the race that they looked like miniature jasmines lying in a neat huddle on the floor. In tight tuffets and mats it nestled to the ground ... this little beauty proves not only a new species, but the type of a new race, cousin to Daphne and Stellera but distinguished now as Farreria.' At Edinburgh it was named *Farreria pretiosa* but that name has never been validly published with a botanical description and the specimen *Farrer 19* cited by Farrer is not at Edinburgh (fide I.C. Hedge in litt.). Domke, who saw no material, supposed it be a species of *Wikstroemia* allied to *W. myrtilloides*, which is possible, but the species seen by Farrer was probably *W. chamaedaphne* Meissner, known to occur in southern Kansu.

NOTE

For an account of the difficulty of rendering the diversity of Chinese pronunciations into Roman characters, the Wade Giles system of romanization and the modern official Pinyin system now superseding this, see W. T. Stearn, 'Chinese puzzle' in *The Garden* 116, 85–89 (1991).

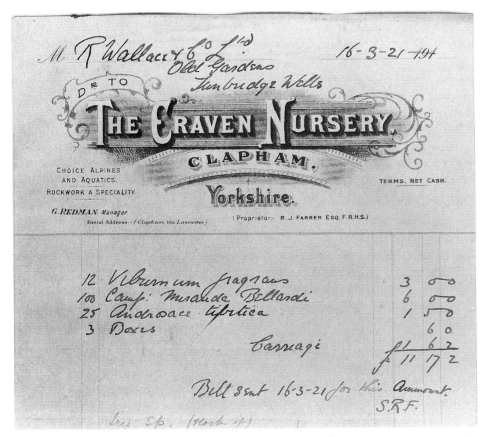

14. Letterhead of the Craven Nursery showing *Viburnum farreri* (*Viburnum fragrans*) for sale in 1921 at five shillings

8

THE HERBARIUM COLLECTIONS OF REGINALD FARRER

Alan P. Bennell & Jennifer Lamond

The achievements of the plant collectors are usually appraised by surveying the quality and diversity of plants with which they have enriched our gardens. Reginald John Farrer's explorations in the montane regions of the Kansu-Tibetan border (1914–1915) and the Burmese-Chinese border (1919–20) led to the introduction of a range of elite rock garden plants and shrubs that have enthused subsequent generations of professional horticulturists and amateur gardeners. Extensive cultivation of plants that derive ultimately from seed originally collected by Farrer and his collaborators is ample testimony to his work and is fully recounted in *The Plant Introductions of Reginald Farrer* by E.H.M. Cox.[1]

However the successful exploitation of any newly-collected plant material is impossible without first establishing the botanical identity and status of each specimen. This critical work is the province of plant taxonomy – the systematic classification of plants, and the use of this classification to identify and name them. The basic tools of plant classification are herbarium specimens. Thus the astute plant collector will support his gathering of live material (seeds, cuttings etc.) with carefully pressed and dried samples of each plant, chosen to depict the essential vegetative and floral characteristics of the species. Such herbarium specimens are not simply ephemeral objects to be examined and discarded, but provide a permanently preserved collection for future reference. The herbarium with its ordered filing of plant specimens constitutes a dictionary of the plant kingdom, and provides the most efficient way of storing information about the identity of any plant. A Botanic Garden such as Edinburgh maintains over 12,000 plants in cultivation, but its herbarium contains over 2 million specimens, a much more comprehensive reference to the ¼ million plants known to science.

The field botanist on expedition assigns a sequential number to each plant gathered. This number tags the specimen as it is pressed and dried, and is cross-referenced to a field notebook, where observations of the form, habit and habitat of the plant are recorded. In Kansu Farrer travelled with a Kew-trained gardener, William Purdom, and together between April 1914 and November 1915 they noted

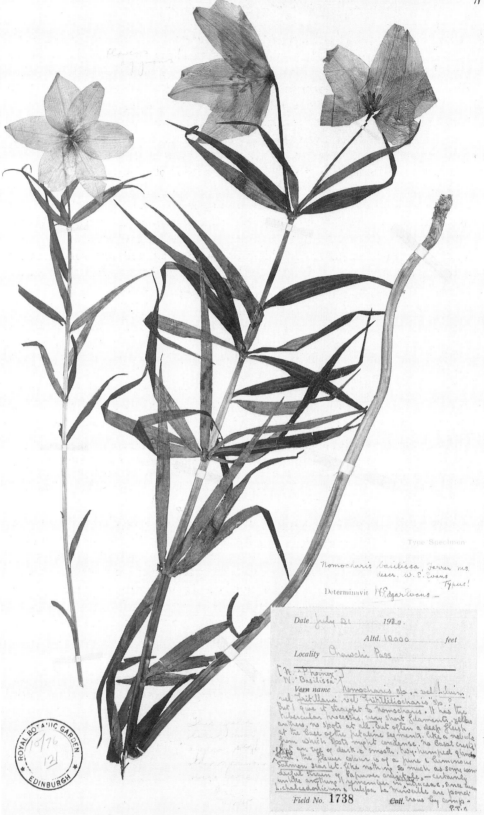

Type Specimen

Nomocharis basilissa, Farrer ms.
descr. W. E. Evans
Typus!

Determinavit *W. Edgar Evans*

Date July 21 1920.

Altd. 12000 feet

Locality Chawchi Pass

[*N. 'Phoenix'*
W. 'Basilissa']

Vern name *Nomocharis sp.,— vel Lilium
vel Fritillaria, vel "Fritillocharis" sp.?
But I give it straight to Nomocharis. It has the
tubercular processes, very short filaments, yellow
anthers, no spots at all, but often a deep flush
at the base of the petaline segments, like a nebula
from which spots might condense. No basal cresset.
Plant an er& dark & smooth. ivy-humped glandular
root. The flower colour is of a pure & luminous
salmon scarlet, like nothing so much as some worn
diluted strain of Papaver orientale,— certainly
unlike anything I remember in Liliaceae. Here even
L. chalcedonicum & Tulipa La Merveille are poor.*

Field No. 1738   Coll. Farrer by comp.
                                  P.T.o

818 collections, most of which are attributed jointly to 'R. Farrer and W. Purdom'. In his subsequent explorations in Upper Burma Farrer was accompanied for the early part of the trip by the notable Perthshire plantsman and explorer E.H.M. Cox. However the 1,112 specimens gathered between April 1919 and September 1920 are all attributed to Farrer alone. His total haul of 1931 collections is modest indeed in comparison to the enormous quantities accumulated by fellow Sino-Himalayan plant hunters such as George Forrest and Frank Kingdon-Ward. Nonetheless, they provide an important if selective catalogue of the plant life of the regions explored, and were the basis for the classification and naming of a number of species.

Edinburgh's pre-eminent position in the study of Sino-Himalayan plants, under the direction of the Regius Keeper Sir Isaac Bayley Balfour, undoubtedly led Farrer to direct all his collections there. Specimens arrived in batches and the accompanying notes from Farrer, and those generated by the recipients provide a fascinating insight to the relationship between the plant collector working in the field and the dedicated taxonomists working in the herbarium. It is their mutual collaboration that ultimately broadens our botanical knowledge.

Farrer's specimens arrived in batches comprising up to three different types of plant material. To William Wright Smith, Assistant Keeper at Edinburgh, he wrote in 1919: 'This letter covers a vastly important despatch of seeds, specimens (addressed to the Professor) and wooden boxes of plants (addressed to the curator)'.[2]

The live plants invariably failed to survive the journey. The seed was of course Farrer's prime concern. His particular passion for rock plants led him to seek especially fine specimens which he was desperate to see successfully propagated:

> F.504 *Isopyrum grandiflorum*. One of the loveliest things I have ever seen in my life. Cool shaded crevice of hard limestone cliffs only. Raise with the utmost care, sowing seeds individually round the edges of 3" pots, slightly covered with rich fine soil, and then put straight out into shaded cool rock-crevices as soon as possible, sedulously avoiding breaking up the roots.[3]

The herbarium specimens were a matter of contention. It should be noted that at the season when Farrer was able to collect seed the condition of the overall plant – lacking flowers, and in the case of some herbs, with all vegetative structures in decline, often made it impossible to collect adequate herbarium material. Thus for about a fifth of his 1,931 collecting numbers no herbarium reference specimen exists. In some cases the material is rather limited, however in others a large amount has been collected under the same number. Such copious material provides duplicate

---

15. (Opposite page) The type specimen (Farrer 1738) of *Nomocharis basilissa*, collected 21 July 1920 at 12,000 ft. at Chawchi Pass, and formally described by Edgar Evans at Edinburgh, from Farrer's manuscript notes: "... the flower colour is of a fine and luminous *salmon-scarlet* like nothing so much as some wonderful strain of Papaver orientale ..."

Type Specimen

F. 178.   Primula No 15.   Primula absophila

Banks of very deep moss & woodland decay. only in the very
highest woodland zone of the Thibetan forests round the Bei
Ling. at 11–12000. among the Pyrolas uniflora & rotundifolia. A
most dainty & charming plant. running freely underground. &
forming carpets many yards across. fl. July 21. (prime a little
earlier): seed. mid. October

specimens, sets of which are regularly exchanged between major taxonomic institutions. The core set of specimens is held at Edinburgh. According to *Index Herbariorum (Regnum Vegetabile* Vol. 9: Part II: Lanjouw J. & Stafleu F.A. 1982) Kansu material has also been distributed to Vienna, in Austria and Burmese material to the Arnold Arboretum in America. In the last 30 years Edinburgh has despatched further limited sets to a number of other relevant organisations:

- 347 specs. (Farrer & Purdom, Kansu) to Univ Kyoto, Japan
- 257 specs. (Farrer, Burma) to Rangoon, Burma
- 215 specs. (Farrer & Purdom, Kansu & Farrer, Burma) to Univ. Michigan, USA

and smaller quantities (30–50 specimens only) to Paris, Osaka, and Yerevan (Armenia). In this way Farrer's reference material has become more readily available to taxonomists worldwide. The presence of a partial set of his Burmese collections in Rangoon, is obviously most appropriate.

A variation in quantity of herbarium material is common to all collectors. At times however, staff at Edinburgh raised questions about the quality of the specimens they had been sent. Professor Balfour was concerned to obtain a comprehensive inventory of all the plants to be found in the region, whereas Farrer's interest was to send back plants which he considered to have horticultural potential. Internal correspondence between Balfour and his deputy Wright Smith illustrates their concern. In this memorandum Wright Smith comments on the contents of a letter subsequently sent by Balfour to Farrer:

Regius Keeper,
The comments are severe but fully justified ... If you wish to leaven it somewhat, you could possibly add a note about the seeds, over 80 packets of which will be distributed in a day or so. These have come in very good condition. Farrer understands these better than he does herbarium specimens ...4

By contrast Balfour was at times most appreciative of the material:

Dear Farrer,
... the sets of specimens which you have forwarded have come to me. They are a beautiful set [of Primulas] and, as I expected, and as you will have gathered, the bulk of them are new ... I must say that your collections are first rate. In one way they differ altogether from those of other collectors that have come – you have concentrated on what are likely to be horticultural plants. The specimens are beautifully dried and are sufficient I hope to enable us to say what they are, even in the absence of the fruit which you so much regret.5

---

16. (Opposite page) The type specimen of *Primula alsophile* demonstrates not only Farrer's particular interest in these attractive herbs and his careful selection of material for herbarium collections to display fully the vegetative and floral characters of the species, but also his neat field notes: "F. 187 *Primula No. 15* Banks of very deep moss & woodland decay, only in the very highest woodland zone of the Thibetan forests round the Bei Ling, at 11–12000, among the Pyrolas uniflora and rotundifolia. A most dainty and charming plant, running freely underground, & forming carpets many yards across. Fl. July 21, (prime a little earlier): seed, mid-October".

Type Specimen

Omphalogramma w-aceifum Franch.
Fide H. R. Fletcher in
NOTES ROY. BOT. GARD. EDIN IX 159.
1949.

Omphalogramma
Y Primula viola-paludis, Farrer et Purdom

F. 74   Primula No 6   (description & plate sent)
This noble omphalogramma expands, fully in late summer very thick
& flannelly leaves of deep dusty opaque green with lighter veins
remarkably suggesting those of some fat Viola of the Hirta-group
but lying out on the ground, too heavy for their flushed fleshy
& stalwart peduncles. The capsule is apparently, round, the calyx fringed with
many teeth: the sp. with its absurd throat, is not at all a free
seeder. One rocky shelf, that had been blue with blossom, yielded
only some 7 - 8 seed-stems: Fl. May: seed nearly all gone. Sept. 3.

Farrer himself contemplated the contrasting perspectives on collecting in his correspondence with Sir David Prain, Director of the Royal Botanic Gardens, Kew:

> ... now a to Meconopsis Prattii and Balfour's two different forms. This plant has at last brought home to me the difference between the field botanist and the worker in a herbarium, for the latter, confronted with a series of 50 dried specimens, perhaps tend inevitably to concentrate on differences and multiply species, while the former, straying among uncounted millions of what is obviously a single, but fluctuating species on the fellside, is driven to the other extreme and seeks one reconciling common denominator among the many evanescent divergencies.[6]

Ultimately it is the taxonomic appraisal of the collections, and their subsequent use in systematic studies, both researches into the classification of particular plant groups, and the accumulation of information about the plants of a geographic area for a Flora, that establishes the value of herbarium collections. Much of the initial work on Farrer material at Edinburgh was carried out by Wright Smith, whose notes show that he would meticulously work through each batch of specimens before reporting his findings to Balfour. After careful scrutiny and comparison many specimens would be placed in existing taxa. A 1916 letter from Wright Smith to Balfour illustrates this:

> Kindly make the following corrections in your list of Plantae Farrerianae. No. 57. *Staphylea holocarpa* Hemsl. var rosea, Rehd. & Wils. My material is very scanty but I have seen further specimens of Wilson's variety and am now of the opinion that Farrer's plant is better interned there than described as new.

A number of other collections were established as quite new:

> I have now made descriptions of the following new species, completing what I have of that collection ...

Buddleia Farrer	Sedum Farreri
Buddleia Purdomi	Sedum Purdomi
Lonicera Farreri	Callianthemum Farreri
Allium Purdomi	Aster limitaneous
Corydalis Purdomi	Aster kansuensis
Cypripedium Farreri	Aster Farreri
Dracocephalum Purdomii	Aster sikuensis

> Farreri's and Purdomi's are possibly too prominent but Mr Farrer will have an opportunity to change these before finally printed.[7]

---

17. (Opposite page) The type specimen (Farrer 74; "Primula No 6") of *Omphalogramma vincaeflora*. The photographs above the herbarium specimen show plants flowered at Edinburgh from original Farrer seed. His typically expressive annotation reads: "The noble omphalogramma expands fully in late summer *very* thick and flannelly leaves of deep dusky opaque green with lighter veins remarkably suggesting those of some fat Viola of the hirta group but lying out on the ground, too heavy for their flushed flesy & stalwart peduncles. The capsule is apparently round, the calyx fringed with many teeth: the sp. with its absurd throat, is not at all a free seeder. One rocky shelf, that had been blue with blossom, yielded only some 7–8 seed-stems. Fl. May: seed nearly all gone. Sept. 3."

The descriptions validating such new names and documenting them for future reference by other taxonomists and by horticulturists wishing to grow the plants were usually published in the scientific journal *Notes from the Royal Botanic Garden Edinburgh*, in a series of papers between 1915 and 1930. Besides Wright Smith, a number of specialists identified and classified specimens belonging to particular groups (see the Bibliographical record by W.T. Stearn[8]). It is interesting to note that the somewhat excessive desire to name these new species after their collectors was never corrected. Of course in many cases Farrer already knew the identity of the plants he collected or was keenly aware of the likelihood that they were new.

Overall the Farrer herbarium has made a significant contribution to our knowledge of plants from Kansu and Upper Burma with important additions to selected genera such as *Gentiana*, *Primula* and *Rhododendron*. They constitute an important resource that continues to be researched. Page and Rushforth's first description of *Picea farreri* published in 1980 and listed in W.T. Stearn[8], was based on Farrer's material. The Edinburgh Revision of the genus Rhododendron by David F. Chamberlain[9] also inevitably drew on Farrer's Rhododendron collections.

What distinguishes the specimens above all else is the supremely expressive quality of the accompanying notes. As in his other writings his flamboyant use of language readily conjures up a clear picture of the appearance of the plant, and the circumstances in which it was growing:

F. 139 *Cypripedium sp.*

A most curious little plant, running about with single shoots and forming wide colonies in sunny glades and mossy woodland soil of the forest zone in the enormous gorges behind Siku, at about 8000 feet, & often in company with C. luteum. It has a noxious aromatic scent., & the ample lip is of shining waxy gold, warted and deformed with knobs & whelks & bubuckles (sic) like Bardolph.

In conjunction with Wright Smith, Farrer subsequently named this plant *Cypripedium bardolphianum*! Thus Farrer's neat hand and informative text bring to life each herbarium specimen. The content of such notes, coupled to the emphasis on horticulturally attractive specimens, makes the Farrer Herbarium a delight to examine, and a truly expressive guide to the flora he did so much to bring to our attention.

REFERENCES

1   Euan H.M. Cox (ed.), *The Plant Introductions of Reginald Farrer* (New Flora and Silva, London, 1930).
2   R.B.G. Edinburgh Archives, Letter from W. Wright Smith to Farrer, 9.10.1919.
3   R.B.G. Edinburgh Archives, Letter from Farrer to I. Bayley Balfour, 22.1.1916. (List of Chinese seeds sent from Mr Farrer from China for Professor Balfour.)
4   R.B.G. Edinburgh Archives, Memorandum from W. Wright Smith to I. Bayley Balfour, [undated].

5   R.B.G. Edinburgh Archives, File Herb 9/4/1. Expedition to Kansu, 1914. Letter from I. Bayley Balfour to Farrer, 6.2.1915. [copy.]

6   R.B.G. Kew Archives, Asia Letters v.149, 1909–1928 [132–145 China and Tibet]. Letter 134 from Farrer to Sir David Prain, 26.6.1917.

7   R.B.G. Edinburgh Archives, Letter from W. Wright Smith to I. Bayley Balfour, 1916.

8   See the preceding article by W.T. Stearn, 'Plant Names Commemorating Reginald Farrer: A Bibliographical Record'.

9   David F. Chamberlain, 'A Revision of Rhododendron,' Notes Royal Botanic Garden, Edinb., 39(2) (1982) pp209–486.

9

FARRER AS ILLUSTRATOR:
THE DIARIES OF REGINALD FARRER

Ann Farrer

Farrer began illustrating as a small boy, drawing the plants he found on his solitary botanical rambles in the countryside around Ingleborough. Judging by his feelings recorded in later life it is likely that, even at that early stage, he felt a mixture of enjoyment, self-criticism and satisfaction over his artistic endeavours. His lack of formal training was obviously in some ways a hindrance to him and he spoke to Clarence Elliott of his desire to study under a master. As with all self taught artists he had to approach problems of technique by trial and error.

It would be useful, before discussing the merits of Farrer's work as an illustrator, to clarify exactly what he was setting out to do. His paintings were basically records of plants he found that particularly caught his eye; not made with a view to exact botanical representation, but rather to capturing a feeling for the plant as he had found it growing in its natural habitat. In this he paid great attention to form and colour. However, his paintings were much closer to Farrer's heart than something solely to be recorded could ever be, and they were fired by inspiration rather than any sense of duty. This enthusiasm did not spread to photography. He complains about the chemicals necessary in producing a photograph and he does not speak of his camera and its pictures with the love he shows for his paintbox and paintings:

> July 1st 1915 . . . photographed P.siterica and the white Trollius and then I was seized with spirits to paint this latter − and none so ill.

> July 28th 1915 . . . sallied out on inspiration to sketch a little corner of the village and then I came in and worked up very satisfactorily, feeling so exalted in this soft balmy warmth that I could hardly bear myself for sheer delight.

Sometimes the emotion he poured into his painting prevented Farrer from producing as good a result as he felt he should be able to achieve:

> July 9th 1920 . . . soon we got it (A.primula.) opening and then in full expansion − the most perfectly beautiful big snow white bells imaginable. The rapture of these moments! . . . I was so flustered with delight that I could only sit and chew on the cud of it. After lunch I put out the spp. (I can hardly bear the thought of those bells

being pressed) and set aside chosen ones to be painted tomorrow. And even then I was so dithered with joy that I could settle to no work.

July 10th 1920 . . . (till 4.30) I was painting the vestal Primula, it came out tolerable, but in my enthusiasm I crowded the picture. Afterwards very tired, couldn't work, sat and feebly read and felt dispirited and despairful, with my stock of philosophy running very low.

August 12th 1915. I filled in the afternoon with working up a sketch – overdone and overdone and overdone by the time I was through.

August 17th 1915. An agitated day, I spent the morning re-doing the sketch – rather worse in the end than the first version.

18. Farrer's seed preparation table, probably 1914

It is important to remember that Farrer was not sitting comfortably at a desk in civilized surroundings while carrying out this work. Frequently he worked literally in the field, or in whatever place he was lodging at the time. Inclement weather and loneliness caused great hardship to Farrer and cannot have helped him to paint:

19. Watercolour painting of F1031, *Nomocharis farreri*

July 10th 1920 ... but I do mean to stick this out and not be driven down til my 3 weeks are up. And I, who thought 3 weeks might yield at least 1 or 2 decent days! sanguine fool! But really this is bad luck. However, I've got a tent, and bed, and books, and brushes and a brain, *all* I have to do is go on being patient and concentrate on what I do possess and refusing to think of things like fine days.

It is poignant to note that this entry in his diary was written only 3 months before his death.

Farrer also had to put up with the problems that all painters of flowers have to face, even in the most ideal of conditions. Namely, the tendency of flowers to wilt once they are cut:

May 16th 1920. Collected plants for a painting and took them home, but they turned so flabby that I could do nothing with them, and having failed to find yesterdays clump of Trollius (eaten by goats) did another Primula.

August 5th 1920. I feared it was too late to begin the big Primula before lunch as it would probably flop fatally while I was feeding before I could finish; so with great determination and virtue I set to at 11.30 spent the next hour and a half writing up the field book to date, and then, when I turned to feed and looked at the Primula, to find they'd anyhow all gone flop like candles in summer beneath the unaccustomed sunshine! and so did I too.

August 6th 1920. Though the big Primula was as naughty and tiresome as possible I got it posed despite all its flaggings and faintings and made this time quite a fair portrait of it, really in character.

I make no apology for leaving Farrer to express his own feelings about his work. Through his eyes we can best feel his joy and despair. In his entry for June 2nd we experience with him the full journey of a picture being created right up to the anticlimax once it is completed:

June 2nd 1919 ... I sat down to paint it (the most marvellous and impressive Rhododendron I've ever seen − a gigantic, excellent, with corrugated leaves and great white trumpets stained with yellow inside − a thing alone, by itself *well* worth all the journey up here and everything. And oddly enough I did not enjoy doing so at first ... a first false start − a second, better, splashed and spoilt, then a mizzle, so that umbrella had to be screamed for and held up with one hand while I worked with the other. Then flies and torment and finally a wild dust storm with rain and thunder came raging up so that everything had feverishly to be hauled indoors and the Rhododendron fell over and all the lights and lines etc. were of course now quite out of gear. However, I'd done as much as I could for the day by 5.30 but even then was so excited that I continued strolling in glorious meditation till dark and dinner. But one moral is − only paint when fresh or before the day's toils; as it is I must trench on tomorrow which ought to be wholly a rush of letters and articles for the next day's dâk that I mean to send off.

June 3rd 1919. The rhododendron gave me such a bad night ... I set to however and satisfactorily finished it though it took till after 12.

20. Watercolour painting of *Rhododendron McKenzianum* Nyitadi 1/v/20. Five days later, Farrer wrote to Euan Cox from The Residency, Nyitadi: "Will you send me ... 2 water-colour tubes of Ivory Black, & 2 tubes of French Ultra (for I'm getting ahead famously with art ...)

21. Watercolour painting of *Omphalogramma spp.*, Chawchi Camp 15/vii/20
(compare illustration 17, p60)

It has been possible through Farrer's diaries to establish what he felt about his painting – the highs and lows of success or failure as he saw it. The critical eye sternly viewed his work yet with an element of self-satisfaction that was part of his enthusiasm. He was not seeking to render a detailed botanical illustration or to create a magnificent composition, but rather to capture the character of the plant he had chosen, frequently setting it in its natural habitat. He paid great attention to form and colour to this end. Farrer's paintings were emotional experiences for him and I believe that it is his excitement at coming across the subject concerned, with the atmosphere around it, that comes through his painting and elevates it above a pure record to become a valuable accompaniment to his writing.

NOTES

All quotations are taken from the diaries of Reginald Farrer which are part of the private family archive at Hall Garth, Clapham.
[A copy of the 1919 Burma diary of Euan H.M. Cox is now deposited at the Royal Botanic Garden Library, Edinburgh.]

APPENDIX 1:
KNOWN LOCATIONS OF REGINALD FARRER'S PAINTINGS

Royal Horticultural Society

The Lindley Library holds the major collection of watercolours from the 1914/1915 expedition to Kansu/Tibet. The paintings are numbered 1–31. The flowers are illustrated in their alpine settings. Occasionally there is a rough drawing in a corner of a flower head from a different angle, or a seed capsule. Hand written captions and comments by Farrer can still be read clearly, for example:

> Meconopsis Prattii. Highest stone shingles of the Min S'an Alps and highest grass slopes of the Da-Tung.
> Note: No my dear Lady, it is no good saying your Aunt Matilda grows this in her garden at Balham. She doesn't. It is a new species, introduced for the first time into cultivation by me in 1914. Siku. 30.6.14.

22. Farrer's comments about 'Aunt Matilda' are on the mount
of his painting of *Meconopsis prattii*

Hall Garth, Clapham (private)

This is the largest collection and includes the paintings of 1919/1920 returned to
the family from Upper Burma. Three folders exist, labelled:

   (i)   Rhododendrons   c.40 watercolours
  (ii)   Other Flowers   c.30 watercolours
 (iii) Scenery   c.25 watercolours

The 'Other Flowers' folder contains a 'List and description of R.J.F.'s paintings,
1919–1920,' by C. Graham, 1974. A hand drawn and coloured plan of the proposal
for the development of the garden at Ingleborough Hall, is also included in this
folder.

Alpine Garden Society

Six water colours of flowers are recorded, dated 1914/1915.

10

THE CORRESPONDENCE OF
REGINALD FARRER

John L. Illingworth

We are fortunate that the Farrer family of Reginald's time, like other families of independent means, was constantly on the move. The private archive at Hall Garth, Clapham, home of Dr John and Mrs Joan Farrer, is largely a collection of letters from Reginald to his mother as either he or she moved around from Ingleborough Hall to their London home and on their regular visits to continental Europe or further afield. The loving, perhaps besotted, mother kept her son's letters from early childhood to the last letters from Upper Burma.

The letters of the two major expeditions of 1914–1915 and 1919 are reasonably well known to the public. There are approximately one hundred sent by Reginald from Kansu or Burma. In this survey I do not intend to cover the incidents of the expeditions again, but instead to use the family archive to round out the picture of Reginald, throwing light on his character in such a way as to complement the other contributions.

For health reasons Reginald did not go to Eton as his brother was to do. After a somewhat solitary childhood the companionship of his own generation at Oxford seems to have been an intoxication to him. The c.twenty-five letters of 1898–1901, often on Balliol or Oxford Union notepaper, create a picture foreshadowing the later 'Brideshead' image of Oxford. Reginald was very much part of a set:

> (1899) . . . I have just been dining with the Monkswells. Lady Monkswell is very nice though not so thoroughly captivating as Lady Carnarvon, and as for her son, he is a never failing source of pleasure to me. The one person I can compare him to is his best friend, Aubrey Herbert, and those two are certainly the most entirely loveable people I have ever met and have always been my best friends at Oxford, being quite distinct from the people one cultivates for a term or two, in passing admiration of some more or less imaginary mental gift.

In spite of, (or perhaps because of), the harelip and possible cleft palate, Reginald, all his life, tended to idolise 'golden young men'. [Aubrey is a case in point. He was the son of the 5th Earl of Carnarvon and his second wife, the eldest daughter of Henry Howard of Greystoke Castle, Cumberland. Where Reginald appeared to fail, Aubrey succeeded. He made a good marriage, achieved distinguished war service and became a Member of Parliament.] The pace of Oxford life however had its price for Reginald and his friend:

> (1899) ... Aubrey has come back at last. We had to produce a Doctor's certificate of ill-health, and so went to a physician who sounded him briefly, tapped him, and said he thought Aubrey had better go for a change of air, adding with only apparent irrelevance − It's cover-shooting just now, I suppose ... Aubrey has asked me out to Porto Fino ... If I can I shall go immediately after the New Year, which I am fixed on seeing from the top of Ingleborough ... I may ask a few people in the vac if you have no objection. We must have the house crowded to be pleasant and all & I hope to get plenty of shooting.

The 'shades of the prison house' were closing in on this Oxford set and careers had to be sorted out:

> (1900) The excitement of the beastly exam is making me feel excessively energetic and I really stand in no need of nerve remedies yet awhile, I have now done 8 papers out of the 13, & I think I stand a fairly good chance of a Second.

> (1900) Aubrey has just purchased a baby jackal, which is a most fascinating little beast, like a fluffy collie-pup with a curly tail ... Aubrey has suddenly irritated all his friends unspeakably by volunteering for the front, & proposes to start on the 17th: How he passed the tests is a miracle.

Reginald did not try anything as drastic as this! The archive shows a long standing interest in politics and, when at home, he supported the local Liberal candidate:

> (1905) Our meetings have been very successful: − a *huge* one at Bentham last night, with plenty of booing Tory boys. However, they were mouse-silent while I held forth: although previously some of them had turned off the light at the main: whence a mighty wrath and scandal. But I imagine Clough's election [as Skipton's M.P.] is safer than houses.

While at Oxford Reginald had considered other options besides a Parliamentary career, he commented in 1898: 'I suppose I shouldn't be acceptable at Sandhurst ... − I should like to move about a little.' This is exactly what he did, but wrote home in a less frequent letter to his Father:

> (1902) Another thing. If on my return I am to superintend the affairs of the Nation in the intervals allowed by those of the Craven Nursery, it is imperative that I should belong to a Liberal Club. In fact the Eighty is clearly indicated, as its Committee sends one stump-orating round the Kingdom − a practice of the first importance to me whose strength, if any, consists in a power of words − whether I can feel the pulse of a popular audience as I could that of Varsity men remains to be tried. But I am vulgarly believed to have the makings of a very good speaker − some say, more of an orator than Raymond Asquith.

Reginald did stand for Parliament. He was not offered a safe seat to contest, was not successful and the attempt was not repeated. One letter survives written from the Temperance Hotel in Ashford, Kent where Reginald must have stayed during the campaign:

> (1910) Managed two big speeches with eclat and without giving a sign of cold or weariness. Now, praise be, there are only two more nights (for this evening will probably be a mere ovation) ... My father and brother were a great help: my brother especially has the makings of a really effective speaker, when he has learned to rely less on his notes and more on his personality ... Raymond Asquith comes down tonight.

The 1915 correspondence concludes the sad tale of missed political opportunities. Clough finally stood down as the Skipton M.P. and Reginald nursed hopes for the Farrer family providing a candidate. He even suggested that his Father could strike a bargain along the lines of holding Skipton firm to the Liberal cause for a Peerage! The opportunity was neither given to Reginald nor his family.

It is possible that the conversion of Reginald to Buddhism was too much for the political hierarchy to overlook. His close friend Aubrey Herbert must have shown some anguish on his behalf to elicit this reply by the new convert:

> (1908) The really religious mind is absolutely bound to take refuge in the world of ideas, and rigidly to avoid sight or contact of any professional religions – most especially of those who pretend to his own faith. For religion lives in the heart: any attempt to give it form in ordinance or liturgy or priesthood, immediately brings the holy thing into grip of the everyday world, and in that vulgarizing material grasp the holy thing evaporates at once, leaving only a delusive shell behind. Your objections don't altogether touch me ... By the way, logic does play a part in my religion, if only insofar that I imperatively demand of my religion that at least it should not be violently abhorrent from logic, like theological Christianity. Gautama's system touches me in the bull's eye because it fully satisfies my sense of religious mystery, my love of ethics and ideals, and, above all craving for a reasonable – not a *proved* but a reasonable and coherent view of the scheme of things ...

Reginald called upon his long-suffering parents for financial support to launch a career which would combine travel, writing and a Nursery garden which was also to be largely the product of his travels. *The Garden of Asia* is an interesting example of this. During the Japan journey the book material was gathered but he was also to write home:

> (1903) I am literally dazed and staggered at the possibilities for border plants and especially Alpines. Japan is an absolutely virgin country! The few plants which are occasionally introduced from Japan by Veitch and so forth, with much blowing of trumpets, and at a preposterous price, are not found by them in the wilds of the mountains but in the cabbage gardens of Tokyo and Kyoto. All this I got from Alfred Unger, the great gardener here, who does not know the importance rock gardening has assumed in England, and told me, that as Alpines have no commercial value, no one has ever approached the Alpine Flora of Japan. Now we know what a commercial value they have in England, so *what* a business opening for a person with discernment – and money!

23. The stand of the Craven Nurseries awarded a silver cup by the R.H.S.

Money is the keyword and throughout the correspondence it was never far from the surface. Any letter to Father was almost bound to contain a further request and usually there was a parallel attempt to win the support of Mother:

(1915) You really musn't be grudging and peevish about the poor Nursery!! How *can* one *living* man, singlehanded, cope in two months with about a hundred orders of one year, while at the same time raising the required enormous stock for the next and at the same time still look after a large Rockery − He can't do it. He must have a definite permanent staff of two, and besides somebody must be devised to care for the Rockery as a separate concern. You haven't yet realised that the Craven Nursery is to be a large, serious and moving concern ... why do you seem to think that in order not to be dilettante one must neglect the opportunity of making strenuous money out of what one already knows and likes, and spend weary hours gathering doubtful and meagre profit out of what one neither knows nor likes. This is a silly waste of time and trouble. We will say no more about it...

In fact the first world war, and the demise of customers and gardeners, was to put an end to this argument.

Although the travel and gardening literature of Reginald Farrer is still greatly respected, we tend to forget that Reginald also took himself seriously as a writer of fiction. Does anyone still read the novels, plays and poetry of Reginald Farrer? This may now, deservedly, be a chapter of only passing curiosity. The first world war writings are a neglected area worthy of highlighting. During 1917 and 1918 Farrer worked for John Buchan at the Department of Information. He was sent to France to report back from the battlefields. In the collection is a letter from Buchan praising his work:

> (1917) I have received and read your first four letters. They are all good ... The description of the Somme battlefield is the best thing that has been done on that battlefield by a long way. In fact, wherever you describe you are first class ...

These 'first class' descriptions survive in the publication *The Void of War: Letters from Three Fronts*, published in 1918, but similarly Reginald spared his mother no details of the horror of the battlefields he visited, describing in 1918:

> ... a special jaunt where no civilian has set foot. The land was all one raw wilderness of dense shell holes, filled with scummy water, with bloated carcases still afloat in them, and all the odds and ends of battle still strewn as they fell, over the interminably vast expanse of vile naked mud, with nowhere a blade of green, or sign or smallest promise of returning life, and only an occasional splintered spar, bare and blasted and black, to mark where once there had been some pleasant little farm and its orchard.

Throughout his life, whenever Reginald was far away, he craved news of the gardens of Ingleborough Hall and his special favourite, the gardened Clapham gorge, or Cliff. The large collection of letters from the two great expeditions contain many requests for specific information about the gardens:

> (1919) Surely you must realise that I am far far more interested in my plants and trees than in Philip and Nellie's presence ... couldn't you go a little methodically to work accumulating definite good garden news and get from my new man a monthly list of germinations? The numbers alone are wanted, and you need not even go near the garden yourself – I am inspired to this little wail by the contrast between you and old Pa Jumps [Cox], who, though he doesn't pretend the smallest interest in plants or knowledge of them, yet has sufficient sense and imagination to realise that Jumps will want to hear how our finds are coming along and weekly sends a list of numbers germinated.

From a much earlier period there was friction between the generations about who should manage the gardens.

> (1910) For the ten or twelve parties who are annually shown my Cliff, you have four or five hundred who are allowed up to see the cave: so do leave off the attempt to make a grievance of that! ... You are spoiling my work: making it hideous and ridiculous ... leave me my valley, rocks and cliff: go and stick Polygonums ... elsewhere. You have fourteen thousands acres: leave me two!!

Not unnaturally, the parents did regard the immediate gardens of the Hall as their preserve, but this did not stop Reginald doing his best to reorganise

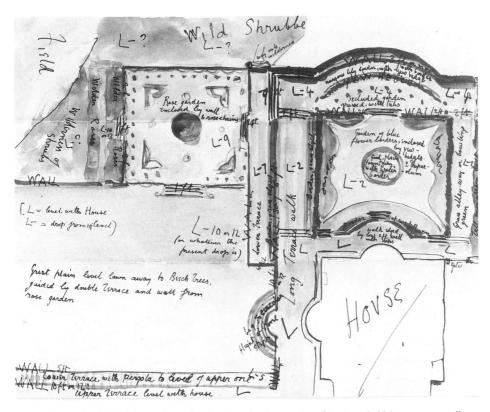

24. Plans for the Ingleborough garden showing the "succession of levels upheld by terraces ..."

things. A Mawson scheme was proposed but the careful parents were never quite ready to implement the grand design:

> (1905) As to the garden-scheme, the essential is a succession of levels upheld by terraces – both for intrinsic beauty, and the dignity of the house frontage. I enclose a rough coloured sketch. The house is so beautifully placed for a fine built garden that one quite longs to see it, but we should have to make money, or be left it, first I think! It is so essential that, even if only one detail, – one walk, is built in a century, that wall or detail must be done in the very best and most solid manner ... I imagine a couple of thousand would turn you out a really lovely garden all complete.

Farrer was not unaware of the changing financial climate for his class, sometimes at the hands of the Liberal politicians. He wrote in 1915: 'Saw Lloyd George and complained of our poverty ... I can't be afforded a chance at Parliament.' To the very end however Reginald retained his faith in the future of the Ingleborough gardens:

(1920) At the same time don't encourage people like the H. Gladstones to come and see it. I am never sure that you properly realise how miserably bad it is still, entirely without design or (natural in the circumstances) upkeep, and a thing still to be very cautious and apologetic about showing either to clever people who know or to rich ones who don't. It is only in about another ten years that the place will gradually become a show-garden, and then only in the sense of a collection, as my labours and expenses of the last fifteen years length mature their fruits. But of course people who on account of my books expect to see a fine garden at Ingleborough simply laugh .. we have as fine a hedge of single Peonies as there is in the Kingdom, probably as good a line of Christmas roses, as good a collection of flag Iris . . . we must go on improving the collection.

This collection of letters reveals Reginald's love of his home and of his mother. In the early stages of the adult correspondence (1903) he wrote to her that they were 'connected like some twins – by a mystic bond of sympathy.' Reginald would have found the post first world war economic and social situation hard to bear, yet I believe he was a realist with an entrepreneurial streak. One of his last letters (1920) is a cry of anguish at 'the way in which everything one ever knew has crumbled.' He felt keenly the dwindling of Liberalism as he saw it into 'an unconvincing amateur – philanthropical pose'. He was tenacious in his defence of the Ingleborough Estate, indeed the attachment of Reginald to his home was set to override all else:

> . . . I find myself believing that all my previous attitudes of mind were nothing but attitudes, with no reality or staying power, or any sort of conviction when it comes to the point of being confronted with the personal loss of all one had . . . And yet, and yet, all these high and complicated thoughts . . . what do they mean more than that I very much (when it comes to the point) dislike having Ingleborough born from me.

NOTES

The family archive at Hall Garth is a private one. Mrs Joan Farrer has arranged the letters as far as possible in chronological order. The editors are grateful to Dr and Mrs Farrer for this exceptional opportunity to quote extensively from the letters. Any subsequent quotation must, of course, receive the consent of Dr and Mrs Farrer.

APPENDIX 2:
MAJOR PUBLIC COLLECTIONS OF LETTERS CONNECTED WITH REGINALD FARRER, HIS EXPEDITIONS AND BOTANICAL INTERESTS

Royal Botanic Garden, Edinburgh. Herbarium Library

Two files designated Farrer cover the plant collecting expeditions:

File Herb/9/4/1   Expedition to Kansu, China, 1914-
File Herb/7/4/1−2 Expedition to Upper Burma, 1919-

Personal files also exist for Purdom, Woodward and Cox.

The Purdom file contains no Farrer correspondence but the Woodward file has correspondence concerning the plants and seeds of the Farrer Purdom expedition, sent by Lt Woodward.

The file for E.H.M. Cox is of great interest. It includes copies of letters, (in the possession of P. Cox of Glendoick), sent by Farrer in 1920 to E.H.M. Cox after the latter's return home from Upper Burma.

The various Directors files are relevant, in particular:

Balfour, I. Bayley Correspondence Farrer

Occasionally there is a copy of the reply sent to Farrer.

Royal Botanic Gardens, Kew. Library

Bound by date and correspondent i.e. Farrer. The replies of the Kew staff are not included. The Archive index had 1 early entry under Purdom (1901/05), no entries under Cox and the following under Farrer:

English Letters 1906/10 vol.115 letter nu. (666−9)
English Letters 1911/20 vol.120 letter nu. (1356−1358)
English Letters 1912/28 vol.126 letter nu. (587A)
Asia Letters   1909/28 vol.149 letter nu. (46−47)

The letters are addressed to curators, Walter Irving, William Watson and William Jackson Bean, and to the Director, Sir Arthur William Hill. The majority of the Asia Letters were sent to Sir David Prain, Director (1905−1922).

APPENDIX 3:
GLASS SLIDES AND PHOTOGRAPHS RELATED TO THE LETTERS

Royal Horticultural Society

The Lindley Library holds on deposit for Hall Garth the 141 glass slides surviving of the 1914/1915 expedition to Kansu/Tibet.

Slides labelled 1−41 of 1914 are mainly views, plus a few of local personalities whose characters are so amusingly filled out in Farrer's publications. Slide 41 is the well known image of William Purdom disguised as a peasant for the expedition to gather seeds for Dipelta floribunda.

The 1915 sequence of 57 glass slides is numbered 61–105, but several bear revised numbers within the group such as 64 and 64a. This is a mixed sequence, largely of views, with some flowers in their natural habitat. Portraits include 87. 'Gogo collecting seed of Isopyrum'; 102. 'Wei Shang ... very proud of new scarlet boots ... at hotel in Chung King.'

Green labels denote a Flowers section, numbered 1–43.

Royal Botanic Garden, Edinburgh

The Purdom holding includes a photograph album dedicated by Farrer to Purdom. The dedication helps to clarify who took these flower photographs in their natural locations:

> To dear Bill, a book of his own photographs from RF 25.2.17. 'I now took a photograph' – works of RF, passim ('I' meaning, invariably, 'you').

The young Cox was likewise to be involved in the photographic work of the Upper Burma expedition.

Hall Garth, Clapham (private)

The family photograph albums contain many excellent examples of the skill of Reginald's father as a photographer. He recorded the development of the Ingleborough Hall gardens and, of course, his family groups include the young Reginald.

Three albums are not the work of John C. Farrer. They cover the expedition of Reginald in 1914–1915. One is labelled 'From Peking to Tibet, 1914–1915'. It includes a frontispiece map and itinerary in R.J.F.'s hand. The second album is labelled 'Lanchow March – Peking Dec', [1915–5D-. Again there is a frontispiece itinerary but also 10 loose photographs, including 'Bill in camp with the boys'. The third album is labelled 'Flowers mostly China', [1914–15–]. Reginald's mother has added a few poignant photographs and comments in the back of the album.

> Reginald's Gentiana Farreri [at Ingleborough Hall]. When the flowers died they failed for lack of his care.

11

THE INGLEBOROUGH ESTATE: HOME OF REGINALD FARRER

Sara Mason

To have preserved an estate for 300 years in one family shows unusual tenacity, especially in the North of England where farming is difficult on marginal land. The Farrer family had lived in Yorkshire for three generations before James Farrer Esq. of Newcastle House, Lincolns Inn Fields built Ingleborough Hall. Their Ingleborough estate was centred on the village of Clapham and the family have been closely connected with Clapham ever since. The majority of land was held in

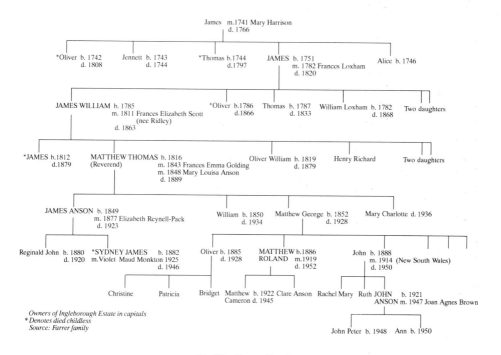

25. The Farrer Family

Clapham-cum-Newby and Austwick, with high land in Ingleton and Horton-in-Ribblesdale parishes. The lordship of the manor of Austwick together with lands in Clapham was acquired in 1782, that of Newby in 1810 and the lordship of Clapham in 1856.

James Farrer invested in land in 1782, the year of his marriage. His elder brother Oliver, called 'Penny Bun' because of his thrifty eating habits on arriving in London, founded the family legal practice there. He too bought land in Clapham for his nephews James William and Oliver. This next generation of brothers was close and held land in the estate, some jointly. Their home was principally in the north. It is said that they also shared a Calvinist Evangelicalism, an attitude which was perpetuated in Clapham.[1]

James and then James William made considerable changes to Clapham in order to landscape the new house. At the top of the valley a lake was made below the road that led to the ancient manor house of Clapdale. A new road cut through the old Archdeacon's Croft in the village and all the land to the east of it was put to the new Hall. Thwaite Lane, or the track to Wharfe, was re-routed by way of tunnels and a grander south entrance was designed from the Settle-Ingleton road. Clearing a suitable space around the Hall involved both rebuilding the vicarage which formerly stood closer to the Church and losing the tithe barn.[2] It appears to have taken some time, between 1807 and 1833, to buy up small parcels of land from several owners to enable the new designs to be carried out. These were not unusual undertakings amongst contemporary landowners: this was a period of such 'improvements', in an age of romanticism. A local architect, John Nicholson of

26. Ingleborough Hall as Farrer knew it

Lawkland, was appointed in 1807 before James died.[3] The house was given a classical frontage between 1820 and 1830 and the offices and stables were built from the old farm buildings at the back in 1842. In 1833 the space in between the house and church was noted to be in chaos, so building must have gone on for some

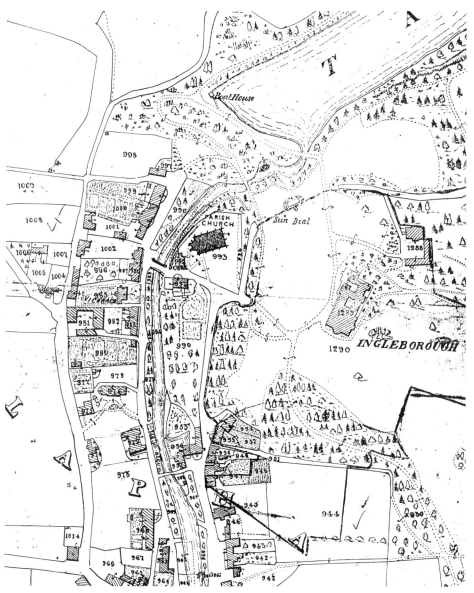

27. Tithe Award 1857, Ingleborough Estate Office

In general, the Ingleborough estate felt less of the fluctuations in the agricultural economy during the nineteenth century than those in other areas of the country. The greater part of the estate was more suited to grass land than arable farming – lyncheted land near Clapham had been enclosed by the nineteenth century – and the value of fields varied according to their nature. Good land near Clapham or immediately adjacent to a relatively low farm was worth more per acre than the thin soil of upland pasture.[5] In the country as a whole wheat prices had fallen due to the free market of the second half of the century and consequently more land was turned over to grass. Prices of sheep and cattle rose between 1850 and 1870 but were somewhat affected by foreign imports between 1875 and 1895. The proximity of the railway at Clapham obviously proved an advantage for transporting dairy products to the towns of Lancashire and Yorkshire and, nearer home, to the railway navvies who were building the Settle-Carlisle line. It does not appear that the Farrers reduced rents during difficult times (as was happening in Westmorland) although some applications were made.[6] From the point of view of the landlord, the revenue from relatively static rents of farms could not keep pace with the costs of running the estate, from the late nineteenth century onwards.

It is difficult to estimate the effect of building up the estate on nineteenth century rural depopulation especially in an area where there was little alternative occupation. In common with other nearby parishes population figures show a general decline in all of Clapham's townships, Clapham, Austwick and Newby apart from the years 1821 and 1831, and 1861 when extra men were probably employed on the estate or on the construction of the railway.[7] From the 1820s some farms were being amalgamated.[8] The sizes of holdings changed considerably between 1851 and 1887:

	1851[9]	1887[10]
Under 5 acres	26	8
5–25	9	6
25–100	10	13
100–200	5	21
200–300		5
300–500	1	1
500+		7
Total	51	61

In the schedule of lands for 1851 the Farrers held 2,550 acres, roughly a quarter of the total land in Clapham which was classed thus: 11,424 acres of which 130 were arable, 4870 meadow or pasture, 400 woodland, 6000 moor and 24 roads. Clapham parish included Austwick until 1879. By 1887 the estate had land in the following townships: 3,296 acres in Clapham-cum-Newby; 1,482 acres in Austwick;

1,070 acres in Lawkland; 1,755 acres in Horton-in-Ribblesdale; 4,160 acres in Ingleton; 11 acres in Thornton-in-Lonsdale, and 816 acres in Dent. Since the land was primarily pastured it would seem that the estate was being consolidated by acquiring land in order to make a ring fence for easier management. On newly enclosed common land gaits or stints mainly for sheep were apportioned to land already held to control the grazing. Enclosure of Clapham's common lands was begun in 1809 by an Act of Parliament and put into effect in 1835.[11] In the award the Farrers held the majority of land but there were over seventy people who were stintholders and had rights of turf-cutting. Other small parcels of common land were enclosed in subsequent years but still the estate includes a very large proportion of unenclosed common land particularly on the steep slopes of Ingleborough.

The Ingleborough Estate was bought and developed as a sporting estate. Averages kept from 1830 over 70 years when the shooting was in hand show regular bags of pheasant, partridge and grouse comparable with other nineteenth century shooting estates in Yorkshire. Ingleborough Moor was retained but the other grouse moors were let from the 1880s. Over 70 years there is an apparent increase in rabbit population. No doubt this caused distress to the grouse shooters since rabbits ate the heather as well as distress to those who were planting trees on the estate.[12] One important improvement to land in the wet North-West was that of drainage, usually done only after a certain amount of land had been amassed. On limestone this was obviously less of a problem than elsewhere though some drainage tiles had been bought from Quernmore Park in 1862.[13]

Much of the Farrers' investment in land during the mid-nineteenth century must have been due not only to the successful legal and banking business but also to the fact that several members of the Farrer family died without children, their legacies benefiting those inheriting the entailed estate. Meticulously kept records of acquisitions and title, together with valuations for death duties which appear to be reasonably accurate, provide a useful running comparison of tenures at given dates. With death duty at 1% if inheritance was direct, and even with 3% to pay on the estate at a brother's inheritance we see a gradual increase in wealth in terms of the annual value of the estate.[14] Added to the natural increase were profits from the sale of land for the building of the railways. In 1847 the estate sold land at Newby to the North Western Railway and the income from this and from the sale of stone and providing clay for bricks was £2881. When the Settle-Carlisle branch line was built by the Midland Railway sales of land at Dent Head between 1869 and 1880 amounted to £2101.[15]

The Farrer family provided access for the visiting tourists who were attracted to this area bordering the Dales. With the coming of the railway visitors were encouraged to stop at Clapham Station where 'The Flying Horseshoe' Hotel had been built in 1850. The tenant of this hotel had also been allowed the tenancy of the Ingleborough Cave first explored by Rev Matthew Thomas Farrer in 1837.[16]

Increasing wealth enabled some philanthropic investment in the village. In 1864, as was happening in other places, the old schoolroom in the churchyard was demolished to make place for the new school which was erected by the family in memory of James William who had died the previous year. James, who inherited from his uncle Oliver in 1866, built the little church at Keasden known as St Matthew's Church in 1873. His brother, Reverend Matthew Thomas Farrer, initiated the refurbishing of St James' Parish Church at Clapham in 1884.[17]

The estate, largely self-financing, was managed by an agent but the family still kept a close interest when in London. The agent dealt with employment on the estate, tenancy agreements and requests for maintenance to farms and cottages as well as capital improvements. The Settled Land Act of 1882 had provided a mechanism for securing estates against possible future depredations. Each successive owner as tenant for life being entitled only to the income, the capital was maintained intact and under trusteeship for the benefit of the estate. Money was reinvested in this way, enabling repairs to be done if rents were sufficient to stand them and improvements made such as the installing of electricity and mains water.

James Anson Farrer, Reginald's father, inherited the estate in 1889 when Reginald was nine years old. He also inherited and continued the tradition of paternalistic involvement in Clapham. In 1890 he restored the old manor house as a reading room, enlarging the parish library from the 100 volumes of 1857 to 2,000.[18] James Anson was a Liberal, and an anti-Imperialist. He published several books on ethnic groups, cultures and religion between 1879 and 1891. As far as the traditional involvement of the squirearchy in local politics was concerned, he himself appears to have played a small but active part, although reluctant to stand himself for Parliament. He resigned the presidency of the Skipton Liberal Association in 1916 in protest against the breakdown of negotiations of peace with Germany. During the last years of the nineteenth century and the early years of the twentieth century social life at Ingleborough included many house parties and visitors, especially during the summer and autumn months. The Sitwells came frequently, especially Edith who visited each year from 1908 until 1914. Violet, Lady Asquith came in 1914. Canon Rawnsley came from Crosthwaite and the Severn family, relatives of Ruskin, from Brantwood. One should not underestimate the cross-fertilisation of ideas about estate management through social contact between the landed gentry. It is perhaps significant that the Farrer's family photograph album and the visitors' book kept from 1889 to 1935 include several neighbouring families.[19] The Lees from Thurland Castle, Tunstall in Lancashire stayed several times although Tunstall was only eight miles away. Lieutenant-Colonel Edward Brown Lees from Oldham had bought Thurland in 1885. In common with other local landowners who were interested in the benefits of hydraulic engineering James Anson explored the possibilities of new-marketed mechanisation to his estate, making use of more consistent power for the water supply from the lake and waterfall.[20] Harry Speight, writing in 1892 about Clapham remarks upon the innovative engineering of the 'Romantic Cascades'. At Clapham early electricity was produced for the Hall and the Church by an overhead cable from a turbine with dynamo supplied by the Kendal firm of Gilbert Gilkes.[21] This was extended to the agent's house, Hall Garth, and to the village from 1904. It was perhaps improvements of this sort and proposals for renovating the Hall at a time of comparative economic stringency due to agricultural depression which prompted the subject of care over the spending of money in correspondence between the family.

A survey of land between 1907 and 1914 notes that which was wet, as at Keasden south of Clapham. Most of the tree planting was done on the less fertile and sheltered of the limestone slopes and on land unsuitable for agriculture. Accounts of receipts and expenditure for timber show an increasing interest in replanting during the last decade of the nineteenth century probably reflecting the interest that both James Anson and Reginald were taking in this aspect of the estate. Between 1861 and 1893 a profit was made, payments to workmen averaging 25% of the gross receipts. Receipts were minimal from 1893 to 1901. From 1901 to 1907 there is expenditure on plantations but no income, offset by substantial income again until 1915 with considerable outlay on replanting. Over a fifteen year period 1901–1915 the net profit realised was 12% or £220.[22]

Reginald's enthusiasm for estate management was never fully tested: he died in Burma in 1920. His brother Sidney inherited the estate from James Anson. Although at its largest in extent at this period, the estate had sold several properties in the Clapham area and bought upland further afield.[23] Matthew Roland, a cousin who had grown up in New Zealand, inherited from Sidney in 1946.

During the Second World War Ingleborough Hall was rented to Captain J.O. Farrer for Stonehouse School which was evacuated from Kent. Post-war legislation caused heavy death duties to the quick succession of owners at this time. In 1952 John Farrer inherited the estate and came from Australia to become, as he says, 'squire on a shoe string'. Finally free of the useful but restrictive Settlement laws[24] he was able to assume total responsibility for the estate and its survival by necessary sales. The Hall had been sold to the West Riding County Council. Land that was sold in 1952 at prices similar to those of a hundred years before, showing very little capital appreciation.[25] The estate is now reduced in size in comparison to the nineteenth century and land has been managed to ensure continuity. John Farrer works on the estate himself. As landowner he is able to exercise control over development to some extent within the terms of Local Authorities. He and his wife Joan run the estate as a small but thriving business, encouraging local industry and settlement as well as farming. With his medical profession to supplement his income Dr Farrer has been able to balance the management of the estate in a traditional manner adjusted to the needs of the present day.

NOTES

NY   denotes North Yorkshire County Record Office, Northallerton
WYAS denotes West Yorkshire Archive Service, Sheepscar Library, Leeds.

REFERENCES

1   Thomas Cecil 2nd Lord Farrer, *Some Farrer Family Memorials 1610–1923*, (Privately published, London, 1923), p.31.
2   WYAS Ingleborough 317. An exchange of land was made with the non-resident vicar of Clapham James Halton in 1807; a Vicarage exchange deed was made in 1833.
3   WYAS Ingleborough 200.
4   Farrer, *Memorials*, p.32.
5   NY ZTW III 5/8. 1907 Newby Moss was worth 1/6d per acre. Tween Rains near Clapham was worth 45/- and Longlands 55/- per acre. Quarry Croft at Nutta was worth 35/- per acre.
6   John D. Marshall and John K. Walton, *The Lakes Counties from 1830*, (Manchester University Press, 1981), p.62.
7   *Victoria History of the Counties of England: Yorkshire*, (3 vols, Archibald Constable, London, 1913), vol 3, p.540. Clapham township. 1801 847 1821 872 1831 944 1841 890 1851 914 1861 809 1871 695 1881 676 1891 712 1901 680 (1989 250).
8   NY ZTW III 4/13. eg 1826 Brownside and Davenanter, 1830 Clapdale, Ingbers and Thornber.
9   Tithe Commutation Schedule: Farrer land. Ingleborough Estate Office.
10 Estate Book: Ingleborough Estate Office.
11 *Act for Inclosing and reducing to stint several Commons and Waste Grounds ... Township and Manor of Austwick ... Parish of Clapham,* 1809. Ingleborough Estate Office.
12 NY ZTW III 12/21. Newby Moor was let in 1882.
13 NY ZTW III 5/8.
14 WYAS Ingleborough 124.
15 NY ZTW III 1/2.
16 Harry Speight *The Craven and North West Yorkshire Highlands*, (E. Stock, London, 1892). The horseshoe, part of the family crest, refers to the possibility of the name Farrer having originated from *farrier*.
17 P.J. Winstone, *A History of the Church in Clapham*, (Privately published, Clapham, 1983), p.64, originally published in *North Yorkshire Journal*, 9 (1982).
18 Winstone, *History of the Church in Clapham*, p.70–71.
19 Visitors' book: Farrer family. Private colleciton.
20 NY ZTW III II/119. Letter Book and Rental.
21 ZTW X11/1 1889. Correspondence with W.T. Goolden & Co of London; ZTW 11/139. Correspondence 1904–1907. The turbine was replaced in 1938.
22 WYAS Ingleborough 124.
23 NY ZTW III 5/8.
24 The Settled Land Act 1886 provided for the suitable use of capital and income from land. That of 1925 made provision for barring the entail.
25 NY ZTW III 1/2.

12

ALPINE PLANTS OF THE INGLEBOROUGH AREA

Jeremy Roberts

Reginald Farrer made his most significant discoveries in far-flung mountain ranges but he had only to walk a short distance from his door in Clapham to find his beloved alpine plants. Indeed, although the variety of true alpines is not great, and the area of 'good rock' is small, the Ingleborough area has been known for its botanical interest since the time of John Ray in the seventeenth century. Even in recent decades there have been significant finds, and I suspect that there are still good plants awaiting discovery.

In the knowledge that the whole of the Dales, even the tops of the peaks, was covered in a vast ice-sheet during the last Ice Age, it is intriguing to speculate on the changes that must have occurred in the area as successive waves of plants invaded from the south, in response to changing climate and soils, over a period of time spanning perhaps 12,000 years. At the risk of over-simplification we may say that our alpines represent the scattered remnants of that first wave of colonists: those plants which by their nature were adapted to survive on the stony, well-drained but mineral-rich soils left by the retreating ice. In neighbouring countries where glaciers still persist, such as Iceland or Norway, we can see many of the same species colonising the gravelly piles of moraine debris just as we imagine they grew in the Dales thousands of years ago.

In the intervening millenia taller, lusher or more vigorous plants have invaded; the soils on gentler slopes have become leached of minerals; and in many areas thick layers of peat have submerged earlier soil-types. The original colonists, with their very particular requirements, have been out-competed and overwhelmed by later arrivals everywhere except in those few places where the rock is limy or steep or high. To see the best of the region's mountain flora aim for those places with all three factors!

A visit to the high cliffs of Ingleborough will soon reveal the essential difference between the influences of limy and non-limy strata on the variety and vigour of the plants. (The cliffs and the slopes above and below them are steep and very loose, and they are not the place for the inexperienced walker, in any weather.) The highest

28. *Anemone nemorosa* at 260m. sheltered in crevice

cliffs are of Millstone Grit and support a very few species – the wavy hair grass (*Deschampsia flexuosa*), bilberry (*Vaccinium myrtillus*), and perhaps the fir clubmoss (*Huperzia selago*). The effects of a couple of centuries of acid-polluted rain from the conurbations to the south has compounded the effects of an acid rock-type. Below them outcrops a good band of the Yoredale limestone, and here one can find mountain plants in great profusion and variety. Every ledge and crack of the limestone supports plants, and the rocks are encrusted with lichens and mosses so that the surfaces are completely masked. The turf below is enriched by the limy stones falling from above, and has a rich covering of thyme, spring sandwort (*Minuartia verna*), lady's mantles, and alpine scurvygrass (*Cochlearia alpina*). These slopes hang above the greatest cliffs on the mountain − steep and rotten beds of sandstone, very short of minerals, and with tolerant plants such as viviparous fescue (*Festuca vivipara*), common male fern (*Dryopteris filix-mas*), and hard fern (*Blechnum spicant*). Nevertheless the best plants of least willow (*Salix herbacea*) grow here, rooted in cracks, cascading down the rocks in mats of contorted branches and tiny rounded leaves. In places, however, the influence of 'flushing' is obvious: one or two ledges are dominated by huge sheets of roseroot (*Rhodiola rosea*) where lime-rich trickles of water enrich the soils with essential minerals.

At the right season, by far the most striking alpine plant of Ingleborough − as also of Pen-y-ghent to the east − is the purple saxifrage (*Saxifraga oppositifolia*). Its season is in fact so early, and varies so much from year to year depending on the season, that one may have to visit the hills on many occasions to see it at its best. If you are lucky you may agree with Raven and Walters[1] that it 'makes as vivid a show here as anywhere in the British Isles'. Although no doubt less abundant than formerly, due to the attentions of collectors, its cushions still straggle down many of the higher cliffs, and in March or April make splashes of purple visible from a hundred yards or more. I know a few remote sections, too, where the saxifrage dares to colonise boulders and screes below the cliffs, and short turf just above them.

The most surprising station for the plant − and very difficult it is to find − is a little patch of scree on Moughton Fell, well away from the high cliffs, where a few dozen rather stunted plants cling on. It would be fascinating to know whether this is a recent arrival here, at hardly over a thousand feet, or whether it has a history dating back to the early post-glacial times. Certainly its close associate in this site, the so-called Teesdale violet (*Viola rupestris*), no doubt has an ancient history here, even though its discovery in the neighbourhood was as recent as 1976. The violet is present, in fact, in many scattered colonies all around the head of Crummackdale, in local abundance in the more open soil and stony areas. It has evaded detection by flowering very sparsely and by mimicking closely the stunted upland form of common dog violet (*Viola riviniana*) which grows in the nearby turf.

Other saxifrages of the high cliffs − and again with a few lower stations as well − are yellow mountain saxifrage (*Saxifraga aizoides*), which grows in sprouts on the limestone cliffs and makes fine yellow flowers later in the season, and mossy saxifrage (*Saxifraga hypnoides*) with cream flowers and lax cushions of three-pronged leaves.

Next in abundance after the purple saxifrage, and usually by far the most conspicuous to visitors in the summer, is the roseroot. This succulent relative of the stonecrops grows in loose patches, its shoots sometimes springing out of tiny cracks in the rock face, or dominating some of the ledges where a little soil has collected. The leaves are a blue- or grey-green, often reddening at the tips, and the yellow flowers make fluffy flat-topped clusters in May and June. The name seems to derive from a rose fragrance of the crushed rootstock.

The alpine saw-wort (*Saussurea alpina*) is an unexpected find on Ingleborough. Its next stations are in the much more suitable shady wet gullies of the Lake District mountains. It has a precarious existence here, growing on one ledge in a ferociously exposed site, where it evaded discovery until 1973.

For the specialist – or the more avid amateur – there are two rare alpine grasses. The first, alpine meadow grass (*Poa alpina*), is restricted to the highest limestone outcrops. It is a robust tufted grass, and unusual in that the Ingleborough form is the 'normal' seed-producing one: this is much rarer in Britain than a 'viviparous' form which produces tiny plantlets instead of seeds. (This propensity crops up in many alpine plants, and seems to be an adaptation allowing easier establishment than the more uncertain production of seed by pollination.) The other 'species', the so-called Balfour's meadow grass, is usually now included in the glaucous meadow grass (*Poa glauca*), although in fact it is also very close to mountain forms of wood meadow grass (*Poa nemoralis*). However it should be classified, there are only a very few plants of it left, on the steepest cliffs.

Other good plants of the high cliffs besides those already mentioned are hoary whitlowgrass – not a grass at all but a crucifer (*Draba incana*) brittle bladder fern (*Cystopteris fragilis*), and green spleenwort (*Asplenium viride*) which almost replaces the familiar maidenhair spleenwort (*Asplenium trichomanes*) at this altitude.

Watching the cliffs over many years brings home the rate at which change occurs in this unstable habitat. Huge cliff falls have occurred, destroying colonies of rare plants, but opening up new rock for colonisation. Populations rise and fall, and feared extinctions have been followed by surprising resurrections. Dr W. A. Sledge of Leeds University told me that one plant, the alpine clubmoss (*Diphasiastrum alpinum*), was widespread on Ingleborough when he first knew the mountain, and yet I know of no recent confirmed sightings, and would be pleased to hear of any. I suspect that the level of some aerial pollutant reached a threshold at some point since then, and the plant succumbed over the whole mountain. Perhaps the same threat hangs over others of the special plants of the Dales . . .

Moist turf in a few places on and near Ingleborough is the site for perhaps the most 'difficult' plant in the area: Dr S.M. Walters named a new species ('microspecies' might be a better term) of lady's mantle which he named *Alchemilla minima*. This is – as you might expect – a tiny plant which may have evolved in response to the close grazing of the turf by sheep. On the subject of lady's mantles, another, *Alchemilla glaucescens*, is very abundant around the Dales on tracksides in thin turf. It is otherwise very local with a few scattered localities in Scotland and Ireland. A third species, *A. glomerulans*, which is a real alpine in Scotland, grows inexplicably in lush marshy turf close to the Ribble above Helwith Bridge. Sorting these three out from the three common Dales lady's mantles – *A. glabra*, *A. filicaulis* subsp. *vestita* and *A. xanthochlora* – is not a job for the faint-hearted!

Having now successfully dragged ourselves away from the high cliffs – and for those who never got that far – there are still a number of good alpine species even at the lower altitudes of the Great Scar limestone, which makes such glorious cliffs around Crummackdale, Chapel-le-dale and Settle, and which has been so catastrophically quarried at Horton. A number of the plants we have already found on the summit cliffs also grow on the lower cliffs, with the addition of a few which for some reason do not reach the upper levels. Alpine cinquefoil (*Potentilla crantzii*) belongs in this latter category, and it is frequent in some of the remaining ash-woods, for instance at Colt Park near Ribblehead. Another candidate is the alpine pennycress (*Thlaspi alpestre*) – but this is best seen in association with lead-mines further east. The best plants are to be sought on the flatter but very exposed tops of the limestone fells such as Moughton Fell, which I have already mentioned in connection with Teesdale violet. The holly fern (*Polystichum lonchitis*) has its English 'headquarters' here, if one can use that term of the few scattered plants which are all that remain. Even in recent years plants have been dug up – now an illegal act – and I once saw two in a rockery in Austwick. The owner said he had searched for them for years!

In a place where spring-waters emerge from the limestone and trickle down broken slopes of impermeable Silurian slate the alpine rush (*Juncus alpinus*) has one of its few English localities. Found here and at Malham in 1976 this rare and difficult plant is otherwise only known from upper Teesdale.

Perhaps the most intriguing of all is the Yorkshire sandwort which was discovered at Ribblehead about a century ago. It was first identified with a Swedish plant called *Arenaria gothica* and because the first known sites were all on tracksides and the plant seemed to be spreading rapidly, it was assumed to be a recent introduction. There are a number of perfectly natural sites now known, and more recent work by Dr G. Halliday of Lancaster University has indicated that, amongst other differences, the local form has the same chromosome number as the Scottish sandwort, *Arenaria norvegica*, but not that of *A. gothica*. It is now regarded as an endemic subspecies (*anglica*) of the former, and distinct from the latter. Unlike its Scottish relative the Ingleborough plant is short-lived, and in order to leave sufficient seed produces its flowers over a much longer period; I have seen it flowering in all months from May to October. The periodically-inundated shallow depressions which form its main habitat are prone to severe drought, to which the sandwort responds by vanishing, only to reappear with equal rapidity after rain. Of all the rarer plants of the mountain, it appeals most to the less energetic: in its best-known station one has to seek it under car-tyres, at least on sunny weekends!

REFERENCES

1   John Raven and Max Walters, *Mountain flowers* (Collins, London, 1956), p.141; see also J.E. Lousley, *Wild flowers of chalk and limestone* (Collins, London, 1950).

13

THE VEGETATION OF THE INGLEBOROUGH LIMESTONE

John S. Rodwell

It was one of Farrer's great gifts that he appreciated the character of particular plants within their landscape context, closely influenced by the rocks and climate which shaped their habitat and growing in intimate association with other species finding the same conditions congenial. It was roaming around Ingleborough, we know, that sharpened Farrer's perception in his early years and, following in his footsteps today, we can inform our own senses and intellect by learning to look in this way at the patchwork of plant communities clothing the scenery he loved so deeply. When Farrer died, the science of ecology was only in its infancy, but that is the name we would give now to this way of seeing, of understanding why plants grow where they do and how they live together in different kinds of vegetation. With its insights, we can build on the inheritance of sensitivity and knowledge which Farrer left us, so as to make his home territory that much more pleasurable and secure for others in the future.

To get some idea of the whole shape of the place, it is best to approach from the west where, along the roads from the Lune valley, we can see the massive platform of Carboniferous Limestone that makes up the bulk of the land mass in the southern Dales, looming up from the Craven lowlands. Above this lies a sequence of softer shales and thinner limestones with a resistant cap of grit that gives such a striking look to the summit of Ingleborough itself and of the neighbouring high hills, but it is the character of the basal limestone that gives much of the scenery and its vegetation their distinctive features. From a distance, the thick beds of pale grey rock can be seen cropping out in huge cliffs (or scars, as they are known locally) and pavements, worn by ice, water and frost in the long centuries during and since the Ice Age. Closer up, we see their surfaces weathered along the prominent bedding planes and joints into a network of hollows and crevices, with boulders and finer rock waste in scree and heaps of surface debris. Perhaps it was the sight of these which gave Farrer his unerring eye for the imitation of nature in the artifice of the rock-garden: certainly, in the wild, these intricately worn exposures offer all manner of surfaces and clefts for colonisation, first by lichens, mosses and liverworts, then by higher plants able to subsist on the thin capping of soil. This soil accumulates only slowly, because the limestone, though it is a hard rock, is almost wholly soluble, even in weakly acid rain water, leaving but little detritus to bulk up the decaying remains of the early invading plants. Very importantly, however, its weathering constantly replenishes the supplies of lime in the soils, because it is composed of virtually pure calcium carbonate. This, together with the natural infertility of the soils and the sharply-draining character of the substrate, have profound effects on the vegetation which develops as colonisation proceeds.

A second very important influence is climate, because the Ingleborough area lies beyond the warm and dry conditions which characterise limestone exposures in the lowland south of Britain, like the Chalk downland of the Home Counties or the Carboniferous Limestone of the Mendips further west. There, summer temperatures often reach 21°C (71°F) and annual rainfall is generally less than 800mm (31″), lower still in the drought-prone eastern part of England. Around Ingleborough, by contrast, mean summer temperatures are a critical couple of degrees C lower at similar altitudes, and reduced further, of course, on moving up into the hills, while rainfall often exceeds 1600mm (62″) per year. This shift to more inhospitable conditions hinders the growth and reproduction of a number of plants that are prominent colonisers of infertile and free-draining limestone soils further south. Thus, it is usually only at lower altitudes and on warmer, south-facing slopes that we would expect to see the pretty horseshoe vetch (*Hippocrepis comosa*), small scabious (*Scabiosa columbaria*) or squinancy wort (*Asperula cynanchica*), while some important woody invaders, like dogwood (*Cornus sanguinea*) and buckthorn (*Rhamnus catharticus*) are also approaching their northern limits hereabouts. Plate 29 shows how the limestone swards in which such warmth-loving plants occur have only just a marginal hold around the Ingleborough area.

For the most part, then, the plant communities of this region have their affinities with the cooler and wetter uplands of north-west Britain and, towards the summits of the hills, conditions can be decidedly montane − cloud-ridden, very windy and quite bitter in winter. It is here, growing on the more inaccessible cliffs and ledges of the thin limestone exposures, that most of the arctic-alpine plants of the area are to be seen, species like the purple and yellow mountain saxifrages (*Saxifraga oppositifolia* and *S. aizoides*) and roseroot (*Rhodiola rosea*) which Mr Roberts' article describes. The vegetation they occur among here, rooted in crevices or precariously festooning the crags, is generally more fragmentary than the damp ledge community that is their home in the Scottish mountains further north (Plate 30), but the survival of these far-flung plants towards the southern limit of their range in Britain is all the more valuable for that.

One further reason for the very great scarcity of some of these species in this area, is that they can only persist out of the reach of grazing animals − and of collectors not so scrupulous as Farrer! Compared with the mountain peaks which drew him

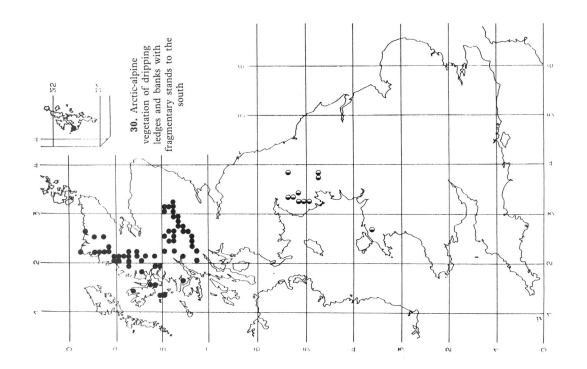

30. Arctic-alpine vegetation of dripping ledges and banks with fragmentary stands to the south

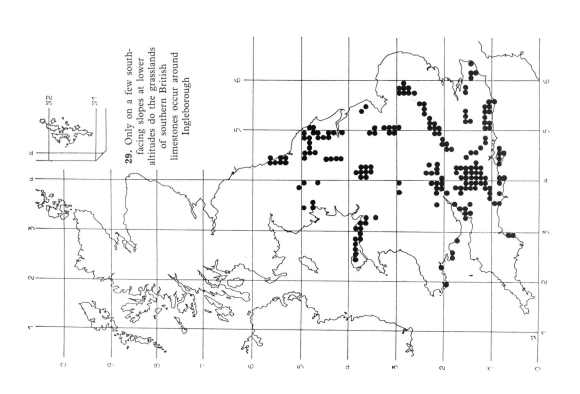

29. Only on a few south-facing slopes at lower altitudes do the grasslands of southern British limestones occur around Ingleborough

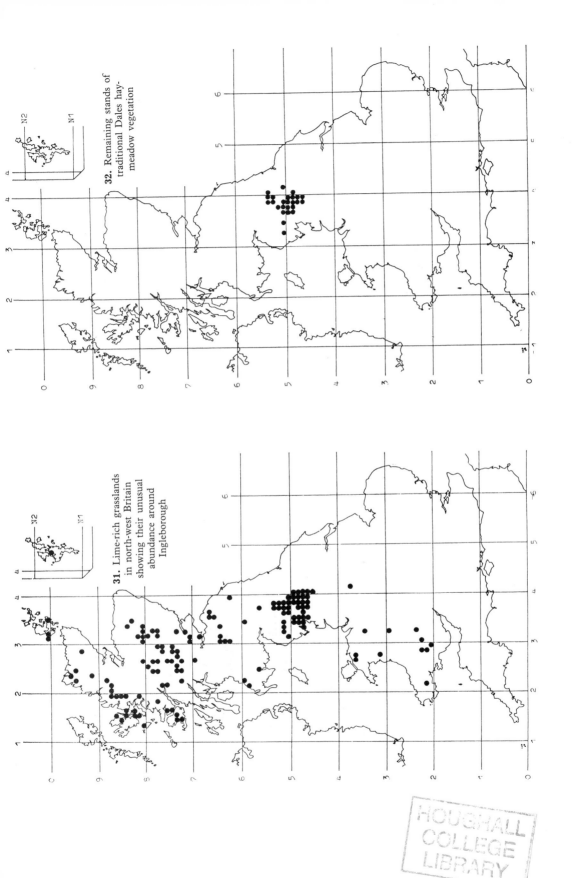

31. Lime-rich grasslands in north-west Britain showing their unusual abundance around Ingleborough

32. Remaining stands of traditional Dales hay-meadow vegetation

abroad, the scenery around Farrer's home is influenced not just by the geology and climate but also, almost everywhere, by man. Indeed, although conditions around Ingleborough are fairly inhospitable, much of its landscape could carry a cover of woodland if invasion of the limestone soils by plants were to run its natural course: the largely treeless state of the scenery, at least at lower altitudes, is artificially maintained. Some stands of more natural woodland can still be found in remote ghylls, where canopies of ash, wych elm, oak, birch, rowan and yew have developed with a rich assemblage of herbs like dog's mercury (*Mercurialis perennis*), bluebell (*Hyacinthoides non-scripta*), primrose (*Primula vulgaris*), meadowsweet (*Filipendula ulmaria*) and the distinctive northern plants, melancholy thistle (*Cirsium helenioides*) and marsh hawksbeard (*Crepis paludosa*). Fragments of such vegetation, the trees and shrubs miniaturised over luxuriant displays of herbs, ferns, mosses and liverworts, can also be seen in the deeper grikes or clefts dissolved down into the limestone pavement, again protected from the predations of grazing animals.

For the most part, the limestone bears the evidence of a long history of forest clearance and pasturing of stock, mostly sheep, though sometimes with cattle also. The result of the assiduous nibbling of these animals on the vegetation which develops on the accumulating limestone soils is thus a fairly short turf, often a delightful mosaic of plants like thyme (*Thymus praecox*), rock rose (*Helianthemum nummularium*), bird's foot trefoil (*Lotus corniculatus*) and Sterner's bedstraw (*Galium sterneri*), among a variety of lime-loving grasses. Swards of this general type occur widely throughout north-west Britain, but they are often decidedly local, marking out small exposures of calcareous rocks or flushing by limy waters. Around Ingleborough, however, we are blessed with great expanses of this sort of grassland (Plate 31), made all the more distinctive by the regional abundance of the blue sesleria grass (*Sesleria caerulea*).

In such close-grazed turf, there is often some influence of seepage of waters draining off the limestone slopes above and keeping the soil surface sodden. This creates a congenial habitat for a whole variety of moisture-loving sedges and dicotyledonous herbs, as well as some distinctive mosses and liverworts. Sheep are often attracted to these flushes, not just by the prospect of a drink, but also because in Spring they provide an early bite of tasty herbage, and a modest amount of trampling from their hooves can help keep the surface open. Then, an especially rich assemblage of plants can develop on the exposed wet silt and stones and around the edges of the broken sods of turf, with some interesting northern species like butterwort (*Pinguicula vulgaris*), lesser clubmoss (*Selaginella selaginoides*), grass of Parnassus (*Parnassia palustris*) and bird's eye primrose (*Primula farinosa*).

Away from such flushed places, the rainfall around Ingleborough is sufficient to induce some surface leaching even in quite shallow limestone soils, so there may often be some herbs in the turf indicative of a loss of lime and a slight rise in surface acidity: the familiar tormentil (*Potentilla erecta*) is a good early indicator of this.

Where superficial deposits like free-draining glacial drift or wind-blown sand or silt were plastered over the landscape as the glaciers retreated, such effects become much more obvious as the controlling influence of the underlying bedrock is muted. Then, more acidic grasslands dominated by fescues, bents and mat grass (*Festuca*, *Agrostis* and *Na⁰dus stricta*) replace the richer limestone swards or, where grazing is not so heavy, some sort of heath. In the period since the Ice Age, much wetter interludes have also encouraged the accumulation of surface humus on the flatter ill-drained ground with a cover of drift, such that we can sometimes find stretches of peat with a thriving bog flora.

33. Ingleborough Hill across the farmland to the west

A final element in the vegetation of the limestone area around Ingleborough is to be seen if we descend into the enclosed fields around the farms of the glaciated valleys that run between the hills. This is the in-by land where the stock are traditionally wintered and from which a hay crop can be taken after the animals have been turned out on to the big enclosures of the limestone slopes or the open hill pastures. Here the soils are deeper than on the rocky ground above, being developed from mixtures of glacial drift, river alluvium and downwash, but in times past they received little enrichment from the farmer apart from a dressing of manure. Under this traditional hay-meadow management, such fields supported a rich herb flora that presented a splendid show when in flower before mowing: the purples and reds of wood cranesbill (*Geranium sylvaticum*), hardheads (*Centaurea nigra*), great burnet (*Sanguisorba officinalis*) and sorrel (*Rumex acetosa*) mixing with the greens, yellows and whites of lady's mantles (*Alchemilla glabra* and *A. xanthochlora*), buttercup (*Ranunculus bulbosus*), rough hawkbit (*Leontodon hispidus*), meadow vetchling (*Lathyrus pratensis*) and pignut (*Conopodium majus*). Now, such vegetation, our nearest approach to the alpine meadows that so delighted Farrer on his journeys through European mountains, is of very restricted occurrence (Plate 32). Most of these fields have been 'improved' for more productive grass crops by the addition of artificial fertilisers and herbicides or by ploughing and reseeding. It is a sad loss, though only one of the ways in which the landscape that Farrer knew has become impoverished. Fortunately, around Ingleborough, there is still much to see and enjoy, but only a continuing commitment to understanding and appreciating its scenery and inhabitants will ensure its survival for our descendants and his.

NOTES

The maps of plant communities in this article have been produced as part of a fifteen year project to describe the vegetation of Britain coordinated by Dr Rodwell in Lancaster University's Unit of Vegetation Science.

CONTRIBUTORS

Alan P. Bennell is a Principal Scientific Officer and Head of Public Services at the Royal Botanic Garden, Edinburgh. He has published papers on Fungal and Higher Plant taxonomy.

Dr. W. Brent Elliott is the Librarian of the Royal Horticultural Society's Lindley Library, and the author of *Victorian Gardens*.

Ann Farrer is a free-lance botanical illustrator who has executed paintings and drawings for a number of publications, including *The Kew Magazine*. She is related to Reginald Farrer, and grew up in the village of Clapham in Yorkshire where he lived.

John Illingworth is an assistant librarian at Lancaster University, with special responsibility for the history collection.

Jennifer Lamond is a Senior Scientific Officer and Associate Curator of the Herbarium at the Royal Botanic Garden, Edinburgh, specialising in the plants of south west Asia.

Audrey le Lièvre is the author of *Miss Willmott of Warley Place*, and contributes to several journals, including *Hortus*.

Prof. John MacKenzie is Dean of Humanities at Lancaster University, and the author of *The Empire of Nature* and editor of *Imperialism and the Natural World*.

Sara Mason is a local historian, with a special interest in the north Lancashire and west Yorkshire area around Clapham.

Dr. Brenda McLean is a fellow in the Geography Department of Liverpool University, with a special interest in Bulley of Ness.

John Roberts is a local botanist whose detailed knowledge of the Ingleborough area has enabled him to make several finds of plants previously unknown in the area.

Dr. John Rodwell is the Director of Lancaster University's Unit of Vegetation Science, and coordinator of the National Vegetation Classification.

Jane Routh is a lecturer in the Department of Visual Arts at Lancaster University.

A. D. Schilling is Kew Garden's Deputy Curator at Wakehurst Place, and contributor to *Plant Hunting for Kew.*

William T. Stearn was Senior Principal Scientific Officer at the Department of Botany, British Museum, and Visiting Professor in Botany and Agricultural Botany at Reading University. He is the author of over 360 publications, including *Botanical Latin.*